SpringerBriefs in History of Science and Technology

More information about this series at http://www.springer.com/series/10085

Alexander S. Blum

Heisenberg's 1958 Weltformel and the Roots of Post-Empirical Physics

Alexander S. Blum
Max-Planck-Institut für
Wissenschaftsgeschichte
Berlin, Germany

ISSN 2211-4564 ISSN 2211-4572 (electronic)
SpringerBriefs in History of Science and Technology
ISBN 978-3-030-20644-4 ISBN 978-3-030-20645-1 (eBook)
https://doi.org/10.1007/978-3-030-20645-1

This Springer imprint is published by the registered company Springer Nature Switzerland AG
The registered company address is: Gewerbestrasse 11, 6330 Cham, Switzerland

Sources[1]

AMPG	Archive of the Max Planck Society, Berlin
DBP	Dieter Brill Papers (in private possession of Dieter Brill, College Park, MD, USA)
GBP	Gregory Breit Papers, Yale University
NBLA	Niels Bohr Library and Archives of the American Institute of Physics, College Park, MD
PJP	Pascual Jordan Papers, Staatsbibliothek zu Berlin
PRP	Paul Rosbaud Papers (in private possession of Vincent Frank, Basel, Switzerland)
PSC III	Pauli Scientific Correspondence 1940–1949 (von Meyenn 1993)
PSC IV-I	Pauli Scientific Correspondence 1950–1952 (von Meyenn 1996)
PSC IV-II	Pauli Scientific Correspondence 1953–1954 (von Meyenn 1999)
PSC IV-III	Pauli Scientific Correspondence 1955–1956 (von Meyenn 2001)
PSC IV-IV	Pauli Scientific Correspondence 1957–1958 (von Meyenn 2005)
WHP	Werner Heisenberg Papers (III. Abteilung, Repositur 93 of the Archive of the Max Planck Society, Berlin)
WZP	Wolfhart Zimmermann Papers (III. Abteilung, Repositur 128 of the Archive of the Max Planck Society, Berlin)

[1] All translations of German sources quoted in the text are by me.

Contents

Chapter 1
Introduction

In 1958, Werner Heisenberg, in his 57th year, jumped the shark. At the Max Planck centennial in Berlin, he presented what others would label his *Weltformel* (World Formula), a final theory reducing all of physics, known and unknown, to the interactions of one elementary quantum field. This caused a substantial media splash, but eventually, especially due to the strong rejection from his old colleague and short-time collaborator Wolfgang Pauli and several other prominent physicists, the physics community concluded that Heisenberg had gone wrong, that he was pursuing a theory whose mathematical consistency was doubtful and which, even if numerical results could unambiguously be extracted from it, could not reproduce the rich particle phenomenology of the subnuclear world.

Why should we care? This book does not aim for a sensationalist story of how the mighty have fallen, of how one of the greatest physicists of the twentieth century lost himself in a blind alley, or of, as Freeman Dyson put it in describing Wolfgang Pauli's presentation of the theory (when Pauli was still in support of it), "the death of a noble animal" (Bernstein 2018, p. 139). Nor is it interested in rehabilitating Heisenberg's approach, as attempted in the historical reconstructions of Heisenberg's disciple Dürr (1982) or in the master's thesis of Rettig (2014). Instead, my aim in this paper is simply to understand how this happened. How could Heisenberg, undoubtedly one of the greatest physicists of the 20th Century, delude himself into believing in a theory that by all accounts was empirically inadequate or even self-contradictory?

There are of course many possible ways to approach this question. A plausible answer might look something like this: Heisenberg was an aging physicist, who had made epoch-making contributions to science as a young man, but whose intellectual and creative capabilities were waning. Yet, his desire for fame and approval had not waned along with them; to the contrary, this character trait had rather been exacerbated by the decline of German science during the Third Reich and after the war and the increasing dominance of the United States. Heisenberg thus urgently hoped and strived for a new epoch-making breakthrough that would put German

A. S. Blum, *Heisenberg's 1958 Weltformel and the Roots of Post-Empirical Physics*, SpringerBriefs in History of Science and Technology,
https://doi.org/10.1007/978-3-030-20645-1_1

physics back onto the map, an epoch-making breakthrough for which he no longer had the capability. This story is certainly not plain wrong, and the reader will find elements of it throughout this book. It is, however, my central thesis that this story of the *Weltformel*, which consists of explanatory categories external to the actual science being done by Heisenberg, is incomplete to the point of being distorting. I will argue that there were in fact essential elements of contemporary physics that made it possible for Heisenberg to pursue his path, fully convinced that he was doing legitimate physics. In particular, it was two features of quantum field theory (QFT), the formal-mathematical language of particle physics, that allowed Heisenberg to lead himself astray.

On the one hand, theorizing within this framework is highly constrained, due to the need to accommodate the complex structures of the underlying physical theories established in the first half of the Twentieth Century, namely special relativity and quantum mechanics. This allowed Heisenberg to start from a fundamental, yet ultimately rather vague, philosophical principle and turn it rather straightforwardly and unambiguously into a specific model for the interactions of elementary particles, without the need for any novel empirical input. I will call the philosophical principle that formed the starting point for Heisenberg's theory *reductive monism*; it is the principle that all elements of the physical world, matter and radiation, can be reduced to one fundamental constituent. No further assumptions about that fundamental constituent need to be supplied, merely the fact that the laws governing its behavior should respect both the special theory of relativity and the central tenets of quantum theory.

On the other hand, once a quantum field theory is constructed according to the established constraints, its assessment was (and to a large extent still is) far from straightforward. It is a popular misconception that the central difficulty in evaluating a speculative physics program such as Heisenberg's is that it makes untestable predictions about the sub-microscopic constitution of the world. Actually, it were far more mundane issues that were impossible to resolve in assessing Heisenberg's theory, namely its mathematical consistency and its empirical adequacy with respect to established empirical data, i.e., its ability to make postdictions.

Just like the ease of constructing models, this resistance to assessment is also a generic feature of QFT, as the theoretical consistency of even the simplest (and empirically highly successful) QFTs, such as quantum electrodynamics, is unproven and was heavily debated in the 1950s; and the empirical consequences of a given QFT can (halfway) unambiguously only be extracted when a perturbation expansion is applicable. In intrinsically non-perturbative theories, such as Heisenberg's, other (and highly questionable) approximation methods had to be used (and were used by Heisenberg), delivering inconclusive results. For this reason, a research program in QFT, once set on its course, was effectively immune to falsification, one's stance toward it defined by beliefs and aesthetic judgments. Heisenberg consequently held on to his non-linear spinor theory until his death in 1976. And also in the German press, the party line was for a long time that the final verdict was still out, a stance most precisely formulated by Heisenberg's compatriot and physicist colleague Pascual Jordan in an article in the *Frankfurter Allgemeine Zeitung* on the occasion of Heisenberg's 70th birthday:

> Elaborating the mathematical consequences of the "*Weltformel*" far enough to make possible a comprehensive comparison with the experimental facts is a very difficult problem. Without a doubt, it is the most difficult mathematical problem ever to be posed in the history of physics. Some specialists even have their doubts whether it can can be solved at all, and in this sense the question of the empirical accuracy of the *Weltformel* is still undecidable [...] Heisenberg himself has not been shaken in his optimism, despite the skepticism of many renowned experts.[1]

It was only around the time of Heisenberg's death in 1976 that one could read, e.g., that "[t]his attempt to combine all subatomic phenomena into a 'unified field theory' might indeed contain mistakes; it still remains sublime",[2] heralding a general consensus even in German media that Heisenberg's attempts had been misguided. Nothing in this book will give reason to question this consensus; but I want to argue that this misguidedness is epistemologically far more interesting than it may seem at first glance.

In fact, it is not just interesting in its own right; I will further argue that the story of Heisenberg's non-linear spinor theory also carries important lessons for contemporary debates on the assessment of theories of quantum gravity, and string theory in particular. This debate began in the mid-2000s when two books by Smolin (2006) and Woit (2006) provided a scathing criticism of string theory, arguing that, despite its meager empirical success, string theory had all but monpolized fundamental physics, forcing young researchers to work in string theory to further their careers, rather than pursue their own out-of-the-box ideas.

In the 2010s, the debate shifted somewhat from sociology to methodology, with Dawid (2013) claiming that string theory was relying on methods of theory confirmation that were non-empirical, yet perfectly scientific, while Hossenfelder (2018) argued that not just string theory but all of particle physics had gone down a methodological blind alley of relying too much on notions of mathematical beauty and elegance as pointers towards true theories. Now there are clear analogies to the Heisenberg story: not merely the grand ambition of constructing the final theory, but more importantly the apparent disconnect with experiment, which has garnered

[1] *Die Aufgabe, die mathematischen Folgerungen der "Weltformel" so weit zu entwickeln, dass umfassende Vergleiche mit den experimentellen Tatsachen möglich werden, ist eine überaus schwierige Aufgabe. Zweifellos ist sie die schwierigste mathematische Aufgabe, die jemals in der Geschichte der Physik gestellt worden ist. Manche Spezialisten hegen Zweifel, ob sie überhaupt in "überzeugender Weise gelöst werden kann, und in diesem Sinne ist die Frage nach der empirischen Richtigkeit der Weltformel heute noch unbeantwortbar [...] Heisenberg selber hat sich trotz mancher Skepsis hervorragender Sachkundiger in seinem Optimismus nicht erschüttern lassen.* Pascual Jordan, "Das Rätsel der Materie" in *Frankfurter Allgmeine Zeitung* of 4 December 1971. An even more positive assessment is to be found in the same day's *Frankfurter Rundschau.* I was able to obtain an overview of the coverage of Heisenberg's theory in the German Press with the help of the Mediendokumentation at the Deutsches Literaturarchiv Marbach.

[2] *Dieser Versuch, alle subatomaren Erscheinungen in einer "einheitlichen Feldtheorie" zu verklammern, könnte tatsächlich Irrtümer enthalten; grandios bliebe er dennoch.* Obituary for Heisenberg in *Der Spiegel* of 9 February 1976, p. 156.

this approach to physics the label "post-empirical", apparently first used by Huggett (2014).[3] But there are equally strong disanalogies, not least of which is the sheer difference in scale, both with regards to explanatory ambition (Heisenberg made only passing reference to quantum gravity) and to the involved manpower. Consequently, in drawing any lessons from Heisenberg's non-linear spinor theory for the contemporary search for a final theory one carefully needs to spell out which analogical elements those lessons are based on, and I will attempt to do this at the end of this book.

At this point, I just want to stress one point concerning the historical perspective this book is supposed to provide on the contemporary debate on string theory and post-empirical physics. By no means do I wish to draw any sort of universal lessons from my historical study. Indeed, I believe that there has been far too much appeal to universal lessons, lessons on how the scientific community should be organized in order to allow new ideas to flourish or on how theories should be assessed in the absence of conclusive experimental findings. In opposition to this, I want to focus on the specific novelties of theoretical physics in the second half of the twentieth century, which is why I have chosen to adopt the "post-empirical physics" label, indicating as it does that we are dealing here with something new: In QFT, we encounter for the first time in the history of physics a paradigmatic theory of fundamental physics whose fundamental equations can be written down with ease but do not have any exact solutions.

Such an historical argument of course requires a far deeper engagement with the nitty-gritty details of Heisenberg's thinking and theorizing than would an argument that purports to extract universal lessons. By far the greatest part of this book is thus devoted to the detailed recounting of Heisenberg's (and later also Pauli's) trajectory from ca. 1945 to 1958. The narrative is divided into three main sections: In Chap. 2, I will outline the construction of the theory, how Heisenberg starting from very general philosophical principles almost inevitably arrived at his theoretical framework. In Chap. 3, I will describe the theory's positive evaluation, Heisenberg's breakthroughs of 1957/58, and Pauli's temporary embrace. In Chap. 4, I will describe the theory's negative evaluation, focusing on Pauli's disillusionment. In the conclusions, I will return to the general points on quantum field theory and theory assessment touched upon in this introduction.

[3] I would like to thank Radin Dardashti for helping me trace the origin of this term, which I have adopted for the title of this book.

Chapter 2
The Origins of Heisenberg's Program

2.1 Philosophical Groundwork

Heisenberg's path of constructing a novel, fundamental theory on the basis of the philosophical principle of reductive monism has its origins in the manifold constraints on scientific research in postwar Germany. In 1941, Heisenberg had become director of the Kaiser Wilhelm (today: Max Planck) Institute for Physics, then located in Berlin.[1] Despite the turmoil of warfare, Heisenberg had a clear, multi-pronged research program, for himself and for his institute.[2] The end of World War II, the German "zero hour", also brought with it an entire reboot of Heisenberg's research program. This was not a matter of choice, a positive decision to start afresh. Rather, it appears to have been the result of frustration: For if we consider the research program that Heisenberg had been pursuing during the World War, we see that Heisenberg was not able to pursue any part of it in the postwar environment for entirely external reasons.

We can identify three pillars of Heisenberg's wartime research program: First, there is obviously the nuclear energy program, which was dominating the research at the institute during wartime. But that was not all that was going on. Another central focus was research on cosmic rays, which formed the subject matter of the institute's colloquium in the academic year 1941/42, resulting in the publication of an edited volume (Heisenberg 1943b). Cosmic rays had been central to Heisenberg's thinking about fundamental physics since 1936, and even though he was well aware of the promise held by the newly developing accelerator technology, he was specifically interested in very rare very-high-energy events (explosions), which one could only observe in cosmic rays.[3] And finally there was his personal work on the theoretical

[1]For detailed biographical information on Heisenberg I rely on (Cassidy 1992) throughout.

[2]On the history of Heisenberg's institute see Kant (1996), Rechenberg (1996), Henning and Kazemi (2016).

[3]See Cassidy (1981), Blum (2017).

A. S. Blum, *Heisenberg's 1958 Weltformel and the Roots of Post-Empirical Physics*, SpringerBriefs in History of Science and Technology,
https://doi.org/10.1007/978-3-030-20645-1_2

and mathematical foundations of quantum (field) theory, which resulted in the highly influential series of S-Matrix papers, written in the years 1942–1944.[4]

The end of the war saw all three pillars of Heisenberg's envisioned research program knocked down, for various reasons. During his postwar internment in Farm Hall, England, Heisenberg had expressed his desire to do further research in nuclear physics, as remarked by the supervisor of the Farm Hall captives, Major T. H. Rittner, in his report of 23 August to 6 September 1945:

> Heisenberg made it quite clear to me that he wishes to continue work on uranium, although he realises that this could only be done under Allied control. (Bernstein 1996, p. 238)

But already at Farm Hall Heisenberg anticipated that there would not be allied control but only allied restrictions and that pursuing *nuclear physics* would be impossible for *political* reasons, as remarked in a secretly recorded conversation with Carl Friedrich von Weizsäcker, Karl Wirtz, and Paul Harteck on 15 September 1945:

> I am a little afraid that the easiest solution for the English would be to say: "They must not do nuclear physics. They may work on cosmic rays." [...] Such a solution annoys me as it would merely mean that we may go home and they would take no further interest in us. (Bernstein 1996, pp. 268–269)

His expectations turned out to be spot on: Legislation of the Allied Control Council prevented any kind of nuclear research in Germany for a decade after World War II.[5] So what about cosmic rays? Heisenberg rightly believed that this would be possible politically, but that it would not be worthwhile given the equipment at the institute's disposal in post-war Germany, as he remarked later in the morning of September 15 to Otto Hahn:

> I would say, if nuclear physics can be done in a big enough way, or cosmic rays with sufficient means, then I should of course like that. If, however, one has to work with a few shabby proportional counters on subjects which have already been exhausted by the Americans, that is just not worth while. […] I had imagined doing nuclear physics and cosmic ray work in greater style in peace time. To do modern physics in a small way is of no use at all. There is a great temptation now for the Anglo-Saxons to say: "[…] [T] hese big technical things are quite out of question for them." This temptation is great. This means, we are permanently put on ice and can do physics on the Roumanian or Bulgarian scale. […] I don't want to do petty physics. Either, I want to do proper physics or none at all. (Bernstein 1996, pp. 270–271)

He did pursue cosmic ray physics further after the institute was refounded in Göttingen,[6] but mainly to put all of the experimental physicists to work. His heart was not in it, given the small scale and the meagre results to be expected, as he remarked to Patrick Blackett on 8 September:

---

[4]See Rechenberg (1989), Blum (2017).

[5]For more on how the scientists at Heisenberg's institute dealt with these restrictions, e.g., by moving into plasma and astrophysics, see an upcoming book on the history of astrophysics within the Max Planck Society by Luisa Bonolis and Juan Andres Leon.

[6]As witnessed by an edited volume on cosmic rays presenting the work at the institute (Heisenberg 1953a).

> Personally, I would go entirely away from experimental physics. But I have had this institute for about five years and I think that these people in some way depend on me and it is nice to go on. (Bernstein 1996, p. 250)

Cosmic ray physics was thus out due to the lack of *technical resources*. Concerning finally his personal research on the foundations of quantum field theory, Cathryn Carson has stated, based on an interview with the British physicist Brian Pippard, who met Heisenberg on the latter's stays in Cambridge in 1947/48, that Heisenberg abandoned his wartime theoretical work on the S-Matrix because of "intellectual exhaustion" (Carson 1995, p. 425). And indeed Heisenberg's work had hit a snag: He had attempted to construct relativistic quantum physics from an entirely new basis, but realized that the conditions he had to impose on the S-Matrix, the central concept of his new approach, were so restrictive that he could only find one very oversimplified toy example that fulfilled them. What is more, this toy model did not provide for the cosmic ray explosions which were to be the central empirical cornerstone of Heisenberg's approach (Blum 2017). And most importantly: Heisenberg had no-one to discuss his work with. As he had lamented already in a review of Gregor Wentzel's textbook on quantum field theory, which had been published during the war (Heisenberg 1943a), there were no young field theorists in Germany. Already during the war, he had had to travel abroad (Denmark, Switzerland, the Netherlands) to discuss his work.[7] And after the war it took a while to train a new generation. In a summary of the current status of his institute, written for a *Festschrift* celebrating the 70th birthday of Otto Hahn on 8 March 1949, then president of the Max Planck Society, Heisenberg had written:

> And finally it will take another two years until there will be available enough young physicists with university degrees in the age group between 25 and 30, which are just as essential for running a research institute as the big machines.[8]

So Heisenbergs work on *foundations of quantum field theory* was out for *personnel* reasons. In his striving for "proper physics," Heisenberg, for the first time since his dissertation, entirely switched his field of research.

What did he turn to?

He turned to more esoteric problems, where he felt that he had, despite the difficulties outlined above, something substantial to offer. There was the problem of superconductivity, a longstanding problem, which was awaiting a novel clever idea, which Heisenberg, at least for a short time, believed he had in hand.[9] In another case, turbulence, Heisenberg could build on his rare expertise on the subject (from his

---

[7] On these trips, see Rechenberg (1989). Also see Walker (1992) on how with these trips Heisenberg was (perhaps unwittingly) agreeing to act as a cultural goodwill ambassador for Nazi Germany in order to keep up his scientific collaboration with foreign scholars.

[8] *Und schließlich wird es noch etwa 2 Jahre dauern, bis wieder genügend junge Physiker mit abgeschlossener Hochschulausbildung, im Alter zwischen 25 und 30 Jahren zur Verfügung stehen, die für den Betrieb eines Forschungsinstituts ebenso notwendig sind, wie die grossen Apparate.* Die Geschichte der KWG und MPG 1945–1949, Teil I & II; Festschrift für Otto Hahn (1949). A copy is to be found in AMPG, V. Abt. Vc, Rep. 4.

[9] As discussed in Carson (1995, p. 428).

dissertation) and also found an immediate field of application in von Weizsäcker's interest in galaxy formation (Heisenberg and von Weizsäcker 1948). Indeed, von Weizsäcker's influence in the face of the lack of a clear research program from Heisenberg appears to have been quite substantial. In Farm Hall, von Weizsäcker had stated that he was not interested in continuing his work in nuclear physics and would rather "like to [...] lecture on physics at some German University and [...] study cosmology and philosophy." (Bernstein 1996, p. 170) And the main foci of Heisenberg's work (aside from the superconductivity) were turbulence in an astrophysical context as well as what he would later refer to as "general philosophy of elementary particles" (*allgemeine Elementarteilchenphilosophie*).[10]

Weizsäcker's influence was not the only reason behind Heisenberg's turn to philosophy: As Carson (1995) argues (p. 69), it was also due to the public demand for philosophical talks in the years after World War II. As Carson points out, Heisenberg was unwilling to give purely philosophical talks and was eager to suffuse his presentations with more specific physical considerations. What Carson does not work out, however, is how by integrating physical considerations into his philosophical talks, Heisenberg was actively, though not necessarily with foresight, formulating his future research program, i.e., how Heisenberg's popular presentations were to directly influence his later scientific work on the *Weltformel*. It is this aspect which I will outline in detail in the following and which explains Heisenberg's use of the term *Elementarteilchenphilosophie*.

To understand Heisenberg's philosophical trajectory, we need to go back to his first attempts at public philosophizing, which arguably begin with a 1928 talk to philosophers at the University of Leipzig (Heisenberg 1984). Building on his recent groundbreaking work in physics, he attempted to embed it into a philosophically informed grand narrative of the history of science, drawing a line from early Greek atomism to quantum mechanics. In the 1928 lecture, Heisenberg had already mentioned Democritos, though only in passing, placing him at the beginning of a lineage of natural philosophers that continued with Aristotle, Kepler, and Newton. But four years later, in a lecture given to the Saxonian Academy of Sciences, Heisenberg presented a narrative of the history of science that directly connected ancient atomism with modern quantum mechanics (Heisenberg 1933).

We needn't talk much here about the deeper motivation to frame the story in this way—Carson has cogently argued that we witness here Heisenberg's attempt to connect modern science to a German ideal of *Bildung*, which centrally involved an education in the Classics (Carson 2010, e.g., on p. 35).[11] What interests us here, is

[10]Letter to Pauli of 3 February 1950, PSC IV-I.

[11]Heisenberg's father August had taught Classics at secondary school and university, and Pauli, after Heisenberg returned to the atomist narrative in the 1948 lecture to be discussed below, granted: "Heisenberg's relation with antiquity [is] alive and real. It is quite natural that he returns to this subject time and again" (*Heisenbergs Beziehung zur Antike [ist] lebending und echt. Es ist ganz natürlich, dass Heisenberg immer wieder auf dieses Thema zurückkommt.*) Letter from Pauli to Markus Fierz of 17 July 1948, PSC III. Heisenberg (1969, p. 7) himself later chose to place his reading of Plato's *Timaeus* as a high school student at the beginning of his intellectual autobiography.

how exactly Heisenberg established this connection. And here we witness a second, explicitly stated, motivation driving the narrative as presented in 1932, namely to protect science from "one-sidedness and hubris".[12] This motivation[13] was to be balanced with presenting quantum theory as actual scientific progress, an assessment that was far from obvious to more traditionally-minded scientists.[14]

Heisenberg performed this tightrope act in the following manner: The history of science, beginning with the Greek atomists, is a history of epistemological concessions. Realizing this prevents a one-sided view of science and at the same time explains the newest step towards abstraction and loss of determinism as the natural mode of scientific progress. Already Greek atomism had stripped the atoms of most qualities, such as color, taste, smell. These qualities could instead be explained analytically through the geometrical configurations of atoms that themselves did not have these qualities. But this meant that these qualities, their "true nature"[15] at least, were also removed from the purview of the natural sciences. This Greek atomistic program, Heisenberg now argued, had been completed by quantum mechanics, which had also stripped the atoms of their mechanical qualities. Quantum mechanics represented the last great epistemological concession in the historical development of atomism, as he outlined most forcefully in a talk in Göttingen the following year (Heisenberg 1934, p. 10):

> The consistent pursuit of the given scientific path led to contradictions, and the fact that these contradictions could not be resolved despite the physicists' determined attempts soon taught the scientists that they had reached an epistemological abyss, which was challenging the last foundations of exact scientific thought. And this was no surprise. For the atomistic hypothesis contradicts intuitive thought, which teaches us the infinite divisibility of matter. So the existence of indivisible building blocks of matter could only be understood by renouncing intuitive interpretation.[16]

As Heisenberg (1941) further outlined in his famous lecture on the theories of color of Goethe and Newton, it was precisely this fact, that "the smallest building blocks of matter are removed from the domain of our intuitive concepts" (p. 272), i.e., that no further epistemological concessions were possible, that "justifies our assumption that

[12] *Einseitigkeit und Überheblichkeit* (Heisenberg 1933, p. 30).

[13] Which of course strongly rings of the Forman (1971) Thesis.

[14] Two years earlier, in 1930, Heisenberg had held a talk in Vienna where he had addressed such concerns, arguing for the epistemologically satisfactory nature of quantum mechanics while avowing that his listeners might "initially regret that the the intuitive and deterministic classical physics has recently been replaced by an unintuitive physics with statistical laws" (Heisenberg 1931, p. 366).

[15] *Wesen* (Heisenberg 1933, p. 31).

[16] *Die konsequente Verfolgung des vorgeschriebenen wissenschaftlichen Weges führte zu Widersprüchen und der Umstand, dass diese Widersprüche trotz des angestrengten Bemühens der Physiker nicht eliminiert werden konnten, lehrte die Forscher bald, dass sie hier an einen erkenntnistheoretischen Abgrund geraten waren, der die letzten Grundlagen exakt-naturwissenschaftlichen Denkens in Frage stellte. Dies war auch nicht verwunderlich. Denn die Atomhypothese widerspricht dem anschaulichen Denken, das eine unendliche Teilbarkeit der Materie lehrt. Also konnte die Existenz unteilbarer Bausteine der Materie nur unter einem Verzicht auf die anschauliche Deutung verstanden werden.*

the electrons, protons, and neutrons, from which according to contemporary physics all matter is constructed, are really the last indivisible building blocks of matter" (p. 273). Here we find Heisenberg connecting with a second tradition, Goethe's scientific inquiries; Weimar classicism being the second pillar of German *Bildung* next to Greek antiquity.[17] Bringing in Goethe also provided Heisenberg with a reading of the potential end of physics implied by having identified the fundamental constituents of matter and the dynamical laws they obey. It would merely mark the conclusion of modern, objective, analytic Newtonian science as "an exact mastery of nature, removed from the living intuition (*der lebendingen Anschauung entzogen*)" (p. 261), as it had been the destiny of his generation to "walk this path that we have taken to its end." (p. 271) Heisenberg compared this necessity of following the inner logic of modern science to its conclusion with the age of discovery, which could "only find its natural end, when all lands were explored and their riches distributed." (*ibid.*) Similarly, finding a final theory of physics implied "that here, too, the lands to be conquered are not infinite" (*ibid.*), liberating resources to finally pursue further the "Goethean path of observing nature", to achieve a "more alive and more unified attitude towards nature" (p. 274). The final triumph of Newtonian science would thus also mark a new dawn for a subjective, holistic Goethean science.

In his postwar talks, Heisenberg drew on this narrative for the history of science which ended with the establishment of quantum mechanics and the tripartite (electron-proton-neutron) model of the atom. But he added a decisive twist, which is most clearly evidenced in the central text of this period: a popular lecture he gave in the context of a physics conference in Zurich in July 1948 (Heisenberg 1949).[18] Where in 1933 he had stated: "This [atomistic] program has now been, in a certain sense, fully implemented.",[19] he now asserted that the atomistic program was not completed after all:

> But by considering the three elementary substances, i.e., the three kinds of elementary particles—electrons, protons, and neutrons, as the constituents of all matter, we have not entirely completed the program of atomic physics, and this brings me to the actual goal of our current atomic physics.[20]

---

[17]As pointed out by Carson (2010, p. 34). Just like the Greek Classics, the Goethean tradition goes deep with Heisenberg personally. As Yvonne Hütter has pointed out, Goethe's *West-Östlicher Diwan* plays a recurring role in Heisenberg's narrative of his 1925 Helgoland epiphany. See http://www.avbstiftung.de/projekte/artikel/news///die-literarischen-fundamente-von-heisenbergs-unschaerferelation/, last accessed 9 May 2018.

[18]This was an important event for Heisenberg, signaling his return to the international stage and bringing him together with several old friends whom he had not seen for many years, including most notably Wolfgang Pauli.

[19]*Dieses Programm konnte jetzt in einem gewissen Sinne vollständig durchgeführt werden* (Heisenberg 1934, p. 16).

[20]*Aber damit, dass wir jetzt drei Grundstoffe, das heisst drei Sorten von Elementarteilchen—Elektronen, Protonen und Neutronen als die Bestandteile aller Materie ansehen, haben wir das Programm der Atomphysik noch nicht ganz zu Ende geführt, und damit komme ich zu dem eigentlichen Anliegen unserer heutigen Atomphysik* (Heisenberg 1949, p. 95).

There was, Heisenberg argued, one more epistemological concession that had to be made, one final abstraction: reducing the set of "elementary substances" to a single one. In this manner one could draw a line that went back not just to Democritos, but even further, to pre-Socratic monism.

There had indeed been developments in physics since 1932 that contributed to a re-evaluation of Heisenberg's original statement: The discovery of new particles, mainly in cosmic rays (the positron in 1933, the muon in 1937, the pion in 1947), had undermined the simple tripartite model. But even back in 1932, Heisenberg had already known what he explained in 1948, namely that the tripartite model did not include light quanta. The fact that light quanta should be considered a form of matter had been driven home by the discovery of the positron and the apparent possibility of transforming matter (electron-positron pairs) into radiation and back again. But that had already been in the mid-1930s, and it was only now that Heisenberg took this empirical fact and turned it into an indication as to how the atomistic program must be concluded. In his public philosophical reflection, Heisenberg had thus developed a program for how the reductionist progression of science from ancient Greece to modern quantum mechanics was to be completed: Through a monism that reduced both the constituents of matter and the quanta of radiation to one single *Grundstoff*, which Heisenberg would sometimes refer to as "energy" (Heisenberg 1949, p. 97).[21]

The 1948 Zurich lecture was the apex of Heisenberg's postwar *Elementarteilchen-philosophie*. Already at this time, the conditions for Heisenberg's return to formal elementary particle physics were being prepared: The development of renormalized QED in the United States provided the necessary stimulus to alleviate Heisenberg's intellectual exhaustion. And the young field theorists he had so sorely missed were beginning to flock towards his institute. Of the research reports for the Max Planck Institute for Physics, which Heisenberg had to submit regularly to the British military administration, the report of 4 March 1949 was the first one to mention dedicated research on the foundations of quantum field theory,[22] by Heisenberg's Ph.D. student Karl Wildermuth.[23] But the return to quantitative physics did not represent a break in Heisenberg's program. Rather, as we shall see in the following, Heisenberg's research was to a large extent defined by his work on *Elementarteilchenphilosophie*.

[21]The use of this term, I suppose, was meant to conjure up notions of both universality and transformability. It should be noted, however, that in the late 19th Century it was precisely these features of energy that had suggested it as the fundamental concept in a decidedly anti-atomistic worldview, which could remain agnostic about the microscopic constituents of matter. On the history of this theory of energetics see Deltete (1983).

[22]A copy of this report can be found in WHP, Folder 1260.

[23]Heisenberg had met Wildermuth (1921–2005), when Wildermuth was a student at Berlin University during wartime. Wildermuth had been captured by the Soviet army, but fled from captivity, making his way to Göttingen, where he completed his first degree in physics and then became Heisenberg's Ph.D. student (Fässler and Schmid 2006).

2.2 Explicit Formulation

Heisenberg's move from philosophy back to physics occurred in 1949/50, when he became aware of the great advances made in quantum field theory in the USA and Japan. Heisenberg recognized in the new renormalization techniques the possibility of making his monistic philosophy concrete (letter to Pauli of 3 February 1950)[24]:

> Following an old habit, I am sending you the manuscript of a paper on elementary particles [...] The starting point is, however, not quantum electrodynamics, but rather the sort of general philosophy of elementary particles, which I have been practicing occasionally over the last few years. It now appears to me that the program that I had set myself earlier can now be implemented in a mathematically rigorous manner using the the new mathematics of Tomonaga, etc. It thus appears that one is now on mathematically solid ground and that, as Bohr would say,: "man hved, hvad man kan haabe for". [One knows what one can hope for.][25]

Now while Heisenberg's philosophical program was certainly very fundamental, aiming at reducing all of physics to the dynamics of just one constituent, it also seems somewhat vague. In particular, it appears to say nothing at all about the specifics of the dynamical interactions. But, actually, in modern physics after relativity and quantum theory, model construction is so heavily constrained that a simple principle like Heisenberg's can lead to an almost uniquely defined model. This is an essential factor in understanding the post-empirical bent in Heisenberg's theorizing: He could construct the basic features of his model without any recourse to contemporary empirical results and could rely solely on the well-established foundation of the quantum and relativity theories,[26] along with non-rigorous but almost universally accepted appeals to simplicity. The inevitable way in which Heisenberg's Hamiltonian flowed from the combination of his philosophical principle with the basic tenets of modern physics clearly strengthened his belief that he was on the right track: In 1958, he would characterize his theory as describing not the best, but at least the simplest of all possible worlds (Grundgleichung der Materie 1958).

In fact, the reasoning Heisenberg employed will most likely seem familiar to any particle physicist who has constructed a model Lagrangian. If there is just one fundamental constituent from which all known (and unknown) particles are to be constructed, it must have half-integer spin: Integer spin particles, such as the photon, can be constructed from half-integer spin particles (this had a long tradition, the

---

[24]PSC IV-I.

[25]*Einer alten Gewohnheit entsprechend schicke ich Dir das Manuskript einer Arbeit, die von den Elementarteilchen handelt [...] Der Ausgangspunkt ist aber nicht die Quantenelektrodynamik, sondern die Art von allgemeiner Elementarteilchenphilosophie, wie ich sie in den letzten Jahren immer wieder getrieben habe. Nun schien mir, dass man das Programm, das ich mir früher gestellt habe, jetzt mit der neuen Mathematik à la Tomonaga u.s.w. mathematisch sauber durchführen kann. Es sieht also so aus, als stünde man jetzt auf mathematisch festem Grund, und, wie Bohr sagen würde: "man hved, hvad man kan haabe for".*

[26]These foundations are of course saturated with empirical input, an input that has been processed to the point where it resembles purely formal constraints on theorizing.

so-called neutrino theory of light), but not vice versa. The simplest such particle is a spin 1/2 field, a Dirac spinor.[27]

This had farther-reaching consequences than Heisenberg could have envisioned at the time: It pushed him away from the QFT of the time and forced him to pursue his own path. In hindsight the reason for this is rather simple: The simplest interacting theory of spinor fields (without any bosonic fields as force mediators) is a four-Fermi interaction, as first proposed by Fermi in his 1934 theory of β decay. In his first paper on his fundamental spinor ψ theory, Heisenberg thus took the original Fermi interaction as the default assumption for the interaction Hamiltonian of the theory:

$$H = A\overline{\psi}\gamma_{\mu}\psi\overline{\psi}\gamma^{\mu}\psi \tag{2.1}$$

where the coupling constant A in natural units is directly related to the square of a fundamental length l. In fact, Fermi's theory of β decay had been at the origin of Heisenberg's preoccupation with the fundamental length in the mid-1930s. Carson (1995, pp. 462–463) thus claims that Heisenberg's focus on Fermi theory was basically a throwback to his cosmic ray work of the 1930s, but ignores that Heisenberg's postwar one-spinor approach by itself ineluctably led him to a Fermi-type theory.

But Fermi-type theories, despite their long tradition, were squarely outside of the newly emerged framework of renormalized QFT. For most field theories, the methods that had been used to remove the divergences of QED did not work. This was pointed out to Heisenberg by Pauli in a letter of 28 February 1950. And theories with four-fermion interactions fell into the non-renormalizable group, as was explicitly proved by Kamefuchi (1951).[28] So, while the novel methods of renormalized QED had revived Heisenberg's interest in QFT, his philosophical program had pushed him right back out of this framework. This difficulty eventually led Heisenberg to adopt the central novel characteristic of his theory, the indefinite metric in Hilbert Space.

While Heisenberg could not resort to renormalization to remove the usual ultraviolet divergences of relativistic QFT, he could draw on the lessons of renormalization theory concerning the origin of these divergences. As Wüthrich (2013) has rightly pointed out, the process of fixing QED through renormalization relied on making "explicit which features of the theory could be held responsible for its divergence problems" (p. 283). Wüthrich identifies these features rather qualitatively as the elementary process of electrons emitting and absorbing photons. Mathematically, however, this is a perfectly regular process, which also appears in the unproblematic leading-order calculations of the emission and absorption of real photons. The actual divergence difficulties appeared in the propagation of virtual photons, which involves

[27] There is of course the possibility of starting with just a Majorana or a Weyl spinor. While Heisenberg did consider this at times, he ultimately stuck to the well-known Dirac formulation used in the description of the electron. This provided his fundamental spinor with two additional degrees of freedom.

[28] To be quite precise, he proved the existence of divergences that could not be removed by mass and charge divergence, not the existence of an infinite number of primitive divergents, i.e., non-renormalizability in the strict sense.

the vacuum expectation values (VEVs) of products of two field operators at two distinct space-time points. These VEVs give rise to singular functions of these two points (unsurprisingly called singular two-point functions). Mathematically well-defined renormalization was then always preceded by a regularization, the standard method of the time being Pauli-Villars regularization. In such a regularization, the singular two-point functions were explicitly modified (regularized) to remove their singularities. In the resultant finite matrix elements one could then identify well-defined, finite expressions for the radiative corrections to masses and charges, from which one could then determine the renormalization constants. In a final step, the free parameters introduced in regularizing the singular two-point functions could be made infinite again, making the newly introduced renormalization constants formally infinite, but leaving the entire renormalized theory (at least perturbatively) finite.

There had been the hope that regularization might not just be a formal requirement for performing a well-defined renormalization, but might actually be given a physical interpretation, in which case the theory would not involve any more infinities and only a finite and unproblematic renormalization. But those hopes had been dashed by the formal analysis of Pauli and his assistant Felix Villars (Pauli and Villars 1949). To illustrate this point, which is essential for understanding Heisenberg's program, we will focus on a specific VEV, namely that of the commutator of two field operators at distinct space-time points. The way in which the divergences appear here applies *mutatis mutandis* to any other singular two-point function that might appear in a perturbative calculation in QED.

The renormalization of QED in the late 1940s had been done using the interaction picture and perturbation theory. The field operators appearing in the theory were consequently only the free field operators and their covariant commutation relations were well-known since the 1930s. For a free field, the classical field equations are linear in the field and thus indifferent to the non-commuting properties of the quantum field operators. Consequently, the field operators are solutions to the classical field equations, with each linearly independent solution to the classical field equations multiplied by a (non-spacetime-dependent) annihilation or creation operator. This means that also the commutators will be solutions of the classical field equations, albeit singular ones. Indeed, for a spinor field ψ the covariant anticommutator $\{\psi_\rho(x), \psi_\sigma(x')\}$ (which is equal to its VEV in this case, because it is a c-number)[29] is uniquely determined by the demand that it be a solution to the Dirac equation that vanishes for space-like separations between the points x and x' (Pauli's formulation of microcausality), along with some specific demands on the nature of its singularities. It is given by[30]

[29] By "c-number" I mean something that is not an operator, i.e., commutes with all other quantities in the quantum theory.

[30] Compare Schwinger's contemporary magnum opus, the trilogy on QED, in particular (Schwinger 1948, Eq. 2.29) and (Schwinger 1949, Eqs. A.1, A.10, and A.29).

$$
\begin{aligned}
&\left\{\psi_\rho(x), \overline{\psi}_\sigma(x')\right\} \\
&= i\left(\partial\!\!\!/ + m\right)_{\rho\sigma} \Delta(x - x') \\
&= \frac{i}{8\pi^4}\left(\partial\!\!\!/ + m\right)_{\rho\sigma} \epsilon(x - x') P \int d^4k e^{ik\cdot(x-x')} \frac{1}{k^2 - m^2} \\
&= \frac{i}{8\pi^4}\epsilon(x - x') P \int d^4k e^{ik\cdot(x-x')} \frac{(ik\!\!\!/ + m)_{\rho\sigma}}{k^2 - m^2} \\
&\equiv -iS_{\rho\sigma}(x - x') \\
&= \frac{i}{(2\pi)^4}\epsilon(x - x') P \int d^4k e^{ik\cdot(x-x')} S_{\rho\sigma}(k)
\end{aligned} \tag{2.2}
$$

where Δ is the singular two-point function appearing in the simpler commutator of two bosonic scalar fields, P denotes the principal value of the integral, and Schwinger's function $\epsilon(x)$ is ± 1, depending on whether the zero component of the four-vector x is positive or negative. The first equality thus relates the spinorial commutator to the scalar one; for the second equality, the explicit expression for Δ is inserted; for the third equality, the action of the differential Dirac operator is evaluated; the fourth equality is merely a definition of $S(x - x')$, as the fifth one is a definition of $S(k)$, both definitions being introduced here because these two S functions were heavily used by Heisenberg. Note that, despite appearances, $S(k)$, appearing in the last line of the above equation, is not actually the Fourier transform of $S(x - x')$, because of the presence of the ϵ function.

These are just a few of the manifold possibilities for writing the spinorial anti-commutator. The k-integration can in fact be performed, giving Δ or S in terms of Bessel and Hankel functions (see, e.g., the review article by Pauli 1941). One then gets an explicit expression showing the singular behavior of the function in spacetime. It is singular on the light cone, where Δ is given by (Pauli and Villars 1949, Eq. 5):

$$
\Delta(x) = -\frac{1}{2\pi}\epsilon(x)\left[\delta(x^2) + \frac{m^2}{2}\Theta(x^2)\right] \tag{2.3}
$$

and thus has singularities of the δ type, plus a Heaviside discontinuity. S has an additional singularity of the δ' type, because of the derivative acting on Δ.

If one now has several types of virtual particles jointly contributing to some process in QFT through their propagators, then those contributions to the S function simply add up and one has, introducing a spectral mass density $\rho(m^2)$ for the various particles appearing in the theory:

$$
S_{\text{tot}} = \int_0^\infty S_{m^2}\rho(m^2)dm^2 \tag{2.4}
$$

If there is just a discrete number of particles, ρ is just a sum of δ functions for the various mass values of these particles; but a continuous distribution of masses is possible in principle. The essential point made by Pauli and Villars is merely that the singularities arising from the propagation of the various particles simply pile up; in other words, because the anti-commutators of the various free fields must all have the same sign (they need to be positive definite in the limit where $x = x'$, see letter from Pauli to Heisenberg of 28 February 1950), $\rho(m^2)$ is always a positive definite function. In order to regularize S,[31] i.e., in order to remove its divergences, Pauli and Villars thus had to introduce (high-mass) fields with *wrong-sign* anti-commutation relations. Such fields would lead (by applying their creation operators to the vacuum) to states that are not properly normalizable in Hilbert space. The regularized theory thus cannot be given a physical ("realistic") interpretation conforming in particular with the conservation of the norm (i.e., of probability). One thus always (after renormalization) has to undo the regularization and the auxiliary particles have to be removed from the theory again, through a limiting procedure in which their mass goes to infinity. The regularization is in the words of Pauli and Villars purely "formalistic." But they explicitly emphasized that this was not a satisfactory state of affairs, and that the fact that the auxiliary particles behaved almost like regular ones was highly suggestive (p. 436):

> Summarizing, one has to admit that the additional rules which the "formalistic" standpoint has to use [...] could be immediately understood from the "realistic" standpoint and appears as if borrowed from the latter. It seems very likely that the "formalistic" standpoint used in this paper and by other workers can only be a transitional stage of the theory, and that the auxiliary masses will eventually either be eliminated, or the "realistic" standpoint will be so much improved that the theory will not contain any further accidental compensations.

It was this expectation that Heisenberg intended to build on. When we say that the field theories that Heisenberg was forced to work with were non-renormalizable, they were not non-regularizable. If one modified, and this was Heisenberg's proposal, the theory on the level of the fundamental equations, the anti-commutators, and applied to them the same regularization procedure that one usually only applied at the level of the matrix elements, one ended up with a perfectly finite theory, but one in which the auxiliary particles appearing in the Pauli-Villars formalism could not be gotten rid of anymore and would destroy the unitarity and thus the consistency of the theory. Heisenberg's monistic starting point had thereby in a seemingly inescapable chain of reasoning led him to problems with the usual probability interpretation of quantum mechanics.

Heisenberg (1950) originally hoped to circumvent this problem by modifying the time evolution (which was of course responsible for unitarity) through the introduction a non-hermitian Hamiltonian, but was soon convinced, mainly by Fierz (1950), that this would lead to macroscopic violations of causality. But Heisenberg was not to be deterred and around 1952 adopted a novel approach. As explained above, the commutation relations that Heisenberg wanted to regularize were initially those of the interaction representation, in which renormalized QED had originally been formulated. In the early 1950s, mainly in the context of attempts at generalizing the successes of QED to a non-perturbative description, renormalized QFT had been

[31] To be quite precise, Pauli and Villars merely regularized the Δ function itself.

reformulated in the Heisenberg picture. In the Heisenberg picture, the commutation relations of the field operators are no longer simple, divergent solutions of the free field equations. Rather, the Heisenberg picture commutation relations for the field operators involve a solution of the full quantum dynamics and were thus ultimately only accessible through perturbation theory. But Heisenberg hoped that they would still be related to solutions of the classical field equations in a similar manner as for the free fields (where the covariant commutator, the S above, is itself a singular solution of the—free—Dirac equation). He had been able to show that the full non-linear (i.e., interacting) *classical* field equations sometimes had solutions that were less singular than their counterparts for the linear, non-interacting field equations (Heisenberg 1952). Consequently, he hoped that the commutator of the interacting Heisenberg fields might in fact be less singular (or even divergence-free) than the free field commutator $S(x - x')$, and that thus the apparent divergences of his non-renormalizable spinor theory were just an artifact of perturbation theory (Heisenberg 1953b). But this idea he was again disabused of, this time by one of the QFT prodigies in his own institute, Harry Lehmann (1924–1998).

Lehmann had been drafted by the German army after graduating from high school in 1942 and had been captured by the American forces in North Africa.[32] After three years in American captivity, he returned to Germany, where he studied physics, obtaining his Ph.D. with Friedrich Hund in Jena in 1950. While Lehmann's thesis had been on classical electrodynamics, he soon turned to the corresponding quantum theory. When Heisenberg published his first paper on the non-linear spinor program in 1950, Lehmann (probably supported by his advisor Hund, who knew Heisenberg well from his time in Leipzig) wrote a letter to Heisenberg, pointing out a small error, which Heisenberg conceded.[33] Lehmann kept his critical attitude towards Heisenberg's work also after joining Heisenberg's institute as a postdoc in 1952. After meeting Lehmann for the first time, Pauli reported in a letter to Fierz (of 20 June 1955, PSC IV-III):

> Heisenberg desperately wanted to convince me of his theory with rotten mathematics. It was amusing, however, that Mr. Lehmann also does not believe a single word of it. He really made a very good impression on me.[34]

Consequently, Lehmann's work remained largely independent of Heisenberg's program, and during his time in Göttingen Lehmann became one of the fathers of axiomatic QFT, producing several foundational results of lasting importance, such as the LSZ reduction formula or the Källén-Lehmann spectral representation, of which we shall now speak. In Lehmann's work on spectral representations we observe an

---

[32]Most of the biographical information on Lehmann in this paragraph—that which is not drawn directly from archival sources—is taken from Zimmermann (2001).

[33]Letter from Lehmann to Heisenberg of 3 August 1950, letter from Heisenberg to Lehmann of 10 November 1950. WHP, folder 1696, Correspondence L-Z 1950.

[34]*Heisenberg wollte mich unbedingt von seiner Theorie mit fauler Mathematik überzeugen. Amüsant war aber, dass Herr Lehmann ihm auch nichts glaubt. Dieser machte wirklich einen guten Eindruck auf mich.*

interesting dynamics at play: While the Källén-Lehmann representation is an important result in QFT in itself, there is reason to believe that Lehmann wrote the paper in which he derived it (Lehmann 1954) partially as a negative response to Heisenberg's speculations; and indeed Heisenberg cited it as such (Heisenberg et al. 1955, p. 425). What Lehmann obtained was a non-perturbative (exact) expression for the VEV of the covariant (anti-)commutator in the Heisenberg picture (where the commutator is no longer a c-number). We discuss Lehmann's argument for the case of a scalar field—the spinor case introduces some additional difficulties, which are inessential for us here, as the result is essentially the same. Lehmann introduced a complete set of states in Hilbert space, where the states φ_k are labelled by their four-momentum k_μ (there may of course be more than one state belonging to a given four-momentum). Since these are eigenstates of four-momentum (and also include the vacuum, which has four-momentum zero), Lehmann easily determined the x dependence of the matrix elements of the Heisenberg scalar field operators $\phi(x)$ between these states:

$$\langle\varphi_K|\phi(x)|\varphi_k\rangle = \phi_{Kk}e^{i(k-K)x} \tag{2.5}$$

By simply inserting the complete set of states in the expression for the VEV of the commutator, Lehmann then obtained the following result for the interacting fields:

$$\langle[\phi(x), \phi(x')]\rangle_0 = i\int_0^\infty dm^2 \Delta_{m^2}(x - x')\rho(m^2) \tag{2.6}$$

where $\rho(m^2)$ is the spectral density implicitly given by

$$\int_0^\infty \rho(m^2)\delta(k^2 - m^2)dm^2 = (2\pi)^3 \sum \phi_{0k}\phi_{k0} \tag{2.7}$$

where the sum is over all states belonging to some four-momentum eigenvalue. From this equation, one obtains the value of $\rho(m^2)$ at the point $m^2 = k^2$, so one needs to apply this equation for all possible four-momentum eigenvalues to get the full spectral decomposition. Lehmann had thus shown that the VEV of the Heisenberg commutator looked just as in the interaction representation, only that it was not given simply by the properties (masses) of the free fields in the Hamiltonian, but rather determined implicitly by the actual energy spectrum and eigenfunctions of that Hamiltonian, i.e., by actual solutions of the Schrödinger equation rather than merely by parameters to be read off the Hamiltonian. The essential thing was that, since $\phi_{k0} = \phi^*_{0k}$, the spectral mass density ρ was a sum of absolute values and thus still positive definite, thereby shattering Heisenberg's hopes that the full commutator might be less singular than the one for free fields.

With his characteristic optimism, Heisenberg took Lehmann's result as an opportunity, since it allowed him to pinpoint precisely how his theory would have to differ from regular quantum theory: It would have to include (as solutions to the Schrödinger equation) anomalous states for which $\phi_{k0} \neq \phi^*_{0k}$, so that $\phi_{0k}\phi_{k0}$ might be negative. These states were to be seen as part of the Hilbert space, lying, however, in an ill-defined subspace that Heisenberg began referring to as "Hilbert Space II" (Heisenberg et al. 1955). It was to replace the high-energy part of the usual Hilbert Space with a set of anomalous states, which themselves had no well-defined energy and were (just like the regular very-high energy states they replaced) never excited physically, only showing up to regularize sums over complete sets of states as they appeared, e.g., in the Lehmann representation. Given the obvious lack of knowledge of such (or any) exact solutions to the Schrödinger equation, Heisenberg simply postulated an expression for the commutator of his Heisenberg field operators,[35] thereby making strong assumptions as to what the contribution from Hilbert Space II to a sum over a complete set of states would be. He postulated that the Heisenberg commutator would be (at least in first approximation) fully equal to the free commutator, except on the light cone. Here, as we have seen, the free commutator was singular (singularities of δ and δ' type). Heisenberg simply subtracted these singularities.

In order to make the commutator regular on the light cone, Heisenberg took the free commutator and first replaced Δ in Eq. 2.2 by $\Delta_{m^2} - \Delta_0$. This eliminated the δ singularity in Δ. To further eliminate the additional δ singularity that appears in S due to the differentiation of the step function, Heisenberg subtracted from the entire expression the term $\lim_{\eta\to 0} \not{\partial} \frac{m^2}{\eta} \partial_\eta \Delta_\eta$. Here the differentiation eliminates the δ function in Δ, which would otherwise be reintroduced by this second subtraction, while the Dirac operator contains no m term, because that term is not singular (it contains only the discontinuous step function). The modified commutator is thus:

$$
\begin{aligned}
&\left\{\psi_\rho(x), \overline{\psi}_\sigma(x')\right\} \\
&= i\left(\not{\partial} + m\right)_{\rho\sigma}\left(\Delta_{m^2} - \Delta_0\right)(x - x') - i \lim_{\eta\to 0} \frac{m^2}{\eta} \partial_\eta \not{\partial}_{\rho\sigma} \Delta_\eta (x - x') \\
&= \lim_{\eta\to 0} \frac{i}{8\pi^4} \epsilon(x - x') P \int d^4k e^{ik\cdot(x-x')} \left[(i\not{k} + m)_{\rho\sigma}\left(\frac{1}{k^2 - m^2} - \frac{1}{k^2}\right) - \frac{m^2}{\eta}\frac{\partial}{\partial\eta}\frac{i\not{k}_{\rho\sigma}}{k^2 - \eta}\right] \\
&= \frac{i}{8\pi^4} \epsilon(x - x') P \int d^4k e^{ik\cdot(x-x')} \left[(i\not{k} + m)_{\rho\sigma}\left(\frac{1}{k^2 - m^2} - \frac{1}{k^2}\right) + \frac{im^2\not{k}_{\rho\sigma}}{k^4}\right] \qquad (2.8)
\end{aligned}
$$

This then was Heisenberg's expression for his anticommutator, which in time lost its explicitly approximative[36] nature. Also, the above expression is really just the VEV of the anticommutator (which in the interacting theory is no longer equal to the anticommutator itself), another fact that Heisenberg was sometimes rather cavalier about. In any case, we will be encountering this expression again.

---

[35] This was somewhat spuriously motivated by his ongoing assumption that the Heisenberg picture commutators would be solutions to the interacting field equations, which, just like their free counterparts, were vanishing for space-like separated points and singular (though less so) at the origin.

[36] The approximation mainly lay in the assumption that the states in Hilbert Space I only described free fundamental fermions, even though Heisenberg of course needed the existence of many bound states to reproduce the full spectrum of subnuclear particles.

Heisenberg had merely postulated the anomalous states in order to motivate his regularized commutation relations. But around the same time, Pauli and Gunnar Källén had found just such states in their analysis of the Lee Model, a non-relativistic toy QFT, put forward by T. D. Lee in 1954. Lee had presented his model as the first renormalized QFT for which exact solutions could be obtained (Lee 1954), and Pauli and Källén then showed that some of these exact solutions were pathological with a negative norm in Hilbert Space (Källén and Pauli 1955). Christening these states ghosts, Pauli began to suspect that these states might be a generic feature of renormalized QFTs, including QED. Early on, Heisenberg began to identify the states in his Hilbert Space II with the ghosts encountered in the Lee Model.[37] But while the example of the Lee Model made it plausible that such states with negative norm should exist, giving the Hilbert Space an indefinite metric, Heisenberg's second assumption, that these states would remain quarantined in Hilbert Space II was not vindicated. Pauli and Källén had shown that physical transitions to the ghost states were possible in the Lee Model, destroying the unitarity of the S-Matrix. It was to be expected that something similar would happen in Heisenberg's theory. As Pauli put it in a letter to Heisenberg of 18 May 1955: "Why does the ghost stay in the bottle?"[38]

Heisenberg had an idea: Perhaps the Lee Model, where regular states could transition to ghost states, was not representative, and one might have models where such transitions were not possible. In order to make that case (first hinted at in a letter to Pauli of 30 May 1955), he needed to know more about the putative ghost states in his model. After all, he had merely postulated their existence to regularize the commutation relations and had not constructed them as solutions of the Schrödinger equation, as Pauli and Källén had done for the Lee Model. Such an analytic construction was out of the question in Heisenberg's far more complex model, so he took a different route: to infer properties of the ghost state from their postulated contribution to the commutation relations (Heisenberg 1956).[39]

As we have seen, two additional terms were necessary to regularize the commutator (Eq. 2.8):

$$-\frac{i\not{k}+m}{k^2}+\frac{im^2\not{k}}{k^4} \tag{2.9}$$

Following Heisenberg's original heuristic interpretation of his Hilbert Space II (high-mass states giving opposite-sign contributions), one would have expected the correction terms to look something like $-(i\not{k}+M)/k^2-M^2$, i.e., just like the contribution

[37] Letter to Pauli of 10 April 1955, PSC IV-III.

[38] This joke really only works in German, where the same word is used for a genie (in a bottle) and a ghost.

[39] The paper was published on the occasion of the 60th birthday of Friedrich Hund. Heisenberg made the connection to Hund's work, by likening the proposed selection rule that prevented transitions to ghost states to Hund's Rule in atomic physics, which is based on the Pauli exclusion principle and thus on the reduction of the Hilbert space to a physically accessible part (anti-symmetrized wavefunctions as analogue of Hilbert Space I).

from the Hilbert Space I states, only with a large mass M and the wrong sign. However, the first term instead looked like the contribution from a zero mass state (at least the denominator, the numerator curiously contained the mass of the regular state with mass m), an inconsistency first pointed out by Källén in a letter to Heisenberg of 2 March 1953 (PSC IV-II). And Heisenberg agreed that the first term "looks more like mass = zero than mass = infinity", hastening to add that the second term fitted even less into any kind of standard interpretation. In 1956 then, inspired by the Lee Model ghost states, he considered explicitly the low-momentum behavior of the two additional terms, to get a better physical interpretation "in first approximation" (Heisenberg 1956, p. 2). For low momenta, the two terms went as

$$\lim_{\epsilon \to 0} -\frac{i\not{k}m^2}{\epsilon}\left(\frac{1}{k^2} - \frac{1}{k^2 - \epsilon}\right) \tag{2.10}$$

which looks like, questions of the numerator aside, a regular (right-sign) contribution from a mass zero state and a ghost-like contribution from a mass ϵ state. Heisenberg thus concluded that the commutator was regularized by a pair of states, one in Hilbert Space I, one in Hilbert Space II, which necessarily had to appear together, labelling this setup a "dipole between a ghost state and a regular state" (ibid.). Building on this interpretation, Heisenberg made a first attempt at justifying the non-excitability of the ghost states, hoping that it might turn out to be a simple corollary of energy-momentum conservation. Pauli, the originator of the ghost idea, was quite dismissive about Heisenberg's proof (which indeed went nowhere), but could appreciate Heisenberg moving away from the vague language of Hilbert Space II to a more coherent picture:

> It is essential to me that the entire "concept" of Hilbert Space II is replaced by a suitable, well-defined indefinite metric in Hilbert Space, which then allows for an *exact* discussion of the connection [...] (isolation of the anomalous states) that you postulate.[40]

Alongside Pauli's continued but ultimately benevolent criticism, Heisenberg continued to encounter resistance from the young postdocs working on field theory at his institute, collectively known as the *Feldverein* (field association) or by the initials of its founding members LSZ, after the already mentioned Harry Lehmann along with Kurt Symanzik and Wolfhart Zimmermann. The conflict is evident, e.g., in a letter from Zimmermann zu Lehmann, written while Lehmann was staying away from Göttingen with the CERN theory division, then still in Copenhagen. The Japanese physicist Kazuhiko Nishijima had just arrived in Göttingen, when Zimmermann wrote to Symanzik:

> I foresee quite some difficulties between Ni. and Hei. Heisenberg is obsessed with the illusion that Ni. is the man who can understand his theory and work on it actively. Even though I have the flu, I did spend some more time at the institute one day, during Heisenberg's trip

[40] *Wesentlich ist mir aber, dass der ganze "Begriff" Hilbert-Raum II ersetzt wird durch eine passende, wohldefinierte indefinite Metrik im Hilbert-Raum, die es einem dann gestattet, den von Dir behaupteten allgemeinen Zusammenhang [...] (Isolation der anomalen Zustände)* exakt *zu diskutieren.* Letter from Pauli to Heisenberg, 23 June 1955, PSC IV-III.

> to Bonn, to find out more about Ni.'s views (in English!). One can describe them something like this: [...] He has a very low opinion of Hei.'s theory [...] Ni. complained that Hei. had written to him, telling that he would take a week off [...] to work out a program with Ni., which Ni. could then work on [...] Ni. would much prefer to work independently. [41]

And a week later:

> The situation with Nishijima has been further clarified. I was able to ascertain that he came Göttingen only because of LSZ.[42]

Heisenberg ended up not publishing anything with Nishijima. On 20 March 1956, Symanzik wrote to Lehmann and Zimmermann (WZP, Folder Symanzik) that Nishijima had "resisted all the brainwashing." But despite this resistance, Heisenberg soldiered on, eventually proposing a novel explanation for the isolation of the ghost states, which we shall discuss in the next section.

[41]*Zwischen Ni. un Hei. sehe ich einige Schwierigkeiten voraus. Heisenberg hat sich ganz und gar in die Illusion hineingesteigert, Ni. wäre der Mann, der seine Theorie verstehen und aktiv daran mitarbeiten könne. Nun bin ich trotz Grippe einen Tag länger ins Institut gekommen, um während Hei. Reise nach Bonn, die Ansichten von Ni. zu erkunden (auf englisch!). Man kann sie etwa so schildern: [...] Von Hei. Theorie hält er gar nichts [...] Ni. beklagte sich, weil Hei. ihm geschrieben hatte, er würde sich [...] eine Woche Zeit nehmen, um mit Ni. ein Programm auszuarbeiten an dem Ni. [...] arbeiten könne. Ni. möchte lieber selbstständig arbeiten.* Letter from Zimmermann to Lehmann, 30 January 1956, WZP, Folder Lehmann Correspondence.

[42]*Die Lage mit Nishijima hat sich weiter geklärt. Ich konnte erfahren, dass er nur wegen LSZ nach Göttingen gekommen ist.* Letter from Zimmermann to Lehmann, 6 February 1956, WZP, Folder Lehmann Correspondence.

Chapter 3
Heisenberg Triumphant

In this section, I will discuss two key breakthroughs achieved by Heisenberg in the years 1957/58, which ultimately convinced him that he was on the right track, led him to present his theory in several overblown public presentations, and temporarily even convinced Pauli to join Heisenberg in his endeavor. The two breakthroughs concerned the possible mathematical consistency of the theory and its possible empirical adequacy in the face of the burgeoning number of new particles being discovered at high-energy accelerators. We will discuss these two breakthroughs in turn, beginning with the question of consistency.

3.1 Mathematical Consistency

As we have seen, the perceived mathematical inconsistency of Heisenberg's theory was the major point of criticism, both for Pauli and the *Feldverein*. Heisenberg had ultimately left unanswered Pauli's question on how the ghosts in Heisenberg's theory were to be contained. It was only in late 1956 (first mentioned in a letter to Christian Møller of 19 December 1956)[1] that Heisenberg came up with a new idea: While he still had no proof for the unitarity of the S-Matrix in his theory, he believed that he could show that in the Lee Model the ghosts could actually be isolated through a slight modification of the model, a modification which in fact appeared to be closely related to the dipole ghosts he had identified in his non-linear spinor theory. To understand what Heisenberg was doing, we need to make a brief digression and introduce the reader to the Lee Model.

[1] WHP, Folder 1774.

A. S. Blum, *Heisenberg's 1958 Weltformel and the Roots of Post-Empirical Physics*, SpringerBriefs in History of Science and Technology,
https://doi.org/10.1007/978-3-030-20645-1_3

3.1.1 Digression: The Lee Model

The Lee Model (Lee 1954) was modelled on the meson theories of the nuclear interaction that were being pursued at the time, i.e., it described two types of fermionic spin 1/2 nucleons (neutrons and protons, or V and N in the Lee Model) interacting through a bosonic scalar meson (the pion, or θ in the Lee Model). How then, despite this proximity to the nuclear quantum field theories of the day, did the Lee Model achieve its distinction of being the first QFT to have exact solutions?

The known quantum field theories all had conserved quantum numbers, such as, in the easiest and best known case of QED, charge. The usual basis used in these QFTs were the eigenstates of the non-interacting Hamiltonian, which described a certain number of particles, each type of particle having certain values for the conserved quantum numbers; in QED these are the electron (charge $-e$), the positron (charge e) and the photon (charge 0). Solutions of the interacting Schrödinger Equation (or its relativistic generalization) would similarly be characterized by the conserved quantum number taking a certain value. The solutions can thus, a priori, be divided into different sectors, each sector belonging to specific values for all conserved quantum numbers. If now in QED one wanted to construct a solution with a certain charge, say e, one needed to take into account not just a free one-particle electron state, but also states with additional neutral combinations of particles, i.e., with an arbitrary number of photons or electron-positron pairs. A solution of interacting QED would thus have to be an infinite superposition of free particle states, all of which have the same charge.

This was different in the Lee model, where the interaction between nucleons and mesons was limited: The V nucleon could only emit θ mesons (turning into an N nucleon), while the N nucleon could only absorb them (turning into a V nucleon). This implied the existence of two conserved quantum numbers: The number of nucleons (the analog of QED charge) and the number of N nucleons minus the number of θ mesons. Thus all particles in the model, even the θ meson, were charged under some conserved quantum number, and there was no equivalent of the entirely neutral QED photon. In addition, the Lee model was fundamentally non-relativistic. Non-relativistic approximations, where the nucleons were infinitely heavy, were typical features of nuclear QFTs (as in, e.g., the Chew (1954) Model). But the Lee Model took this one step further: Not only were the nucleons infinitely heavy, precluding the production of nucleon-anti-nucleon pairs, also the light mesons themselves had no anti-particles—the corresponding terms, which would have involved the absorption of an anti-θ by a V nucleon and its absorption by an N, were simply dropped from the Hamiltonian in flat-out contradiction with special relativity.[2] Consequently, there

[2] It is, consequently, arguable whether the Lee Model really qualifies as a field theory. In the interaction terms one half of the field operator (corresponding to the inverse process involving anti-mesons) is always dropped and the theory is really only defined in terms of annihilation and creation operators. The honor of being the first exactly solvable QFT would then go to the (massless) Thirring (1958) Model, which was in fact a one-dimensional toy version of Heisenberg's non-linear spinor theory.

was also no equivalent to the neutral particle-anti-particle pairs of QED and for any given value of the two conserved quantum numbers (i.e., for each sector), there was only a finite number of combinations of particles that had to be considered.

Labelling the "free" particle states, i.e., the eigenfunctions of the (trivially solvable) Hamiltonian of the non-interacting theory, as $|n_V, n_N; n_k\rangle$, with n_V, n_N, and n_k the numbers of V nucleons, N nucleons, and θ mesons with momentum k, respectively,[3] a solution with a nucleon number of 1 and a zero value for the second conserved quantum number, could simply be written as a linear combination of just two free states: the state with one free V particle ($|1, 0; 0\rangle$) and the state with one N and θ ($|0, 1; 1_k\rangle$). Lee found one such solution, the physical (dressed, in modern parlance) V nucleon state $|V\rangle$ (defined as the state that goes over to the bare V nucleon for vanishing coupling):

$$|V\rangle = Z\left[|1, 0; 0\rangle + \frac{g}{\sqrt{4\pi}} \int d^3k \frac{1}{\sqrt{2\omega_k}(m_V - m_N - \omega_k)} |0, 1; 1_k\rangle\right] \tag{3.11}$$

where g is the (bare) coupling constant, m_V and m_N are the respective masses of the nucleons, and ω_k is the the energy of a meson with momentum k. Z, finally, is a normalization constant, ensuring that the state $|V\rangle$ has unit norm. If no cutoff is imposed on the momenta of the mesons (i.e., if the integral in the second summand extends over all meson momenta), this normalization constant is infinite, like the (perturbative) renormalization constants in QED. Indeed, if one defined the renormalized, physical coupling constant g_c as in QED, then Z was simply the charge renormalization constant, i.e., $g_c = Zg$. This, however, led to the following expression for Z (obtained simply by taking the norm of the state above)

$$|Z|^2 = 1 - \frac{g_c^2}{4\pi} \int d^3k \frac{1}{2\omega_k(m_V - m_N - \omega_k)^2} \tag{3.12}$$

which is highly problematic, since there is a flagrant contradiction between the positive definiteness of the lefthand side and the fact that the righthand side is negative infinity. This could be resolved, as pointed out by Pauli and Källén by assuming that the free particle states were not normalized to 1—if one normalizes the bare V state $|1, 0; 0\rangle$ to -1 instead, the righthand side of the equation above is multiplied by -1. One could thus postulate an indefinite (i.e., not positive definite) metric in the Hilbert Space of the unobservable bare particles in order to ensure that the physical states (solutions of the interacting Schrödinger Equation) were properly normalized. But while Lee's physical V nucleon state could be saved in this manner, Pauli and Källén discovered another solution of the Lee Model. This solution also reduced to

[3] Note that the numbers of nucleons are not divided into separate momentum states, due to their being very heavy. The "infinite" mass of the nucleons merely meant that their kinetic energy was always considered negligible compared to their rest energy, i.e., to their mass. Each nucleon still gave a finite but constant contribution to the total energy of the state, which was simply the nucleon mass.

a bare V in the non-interacting limit but could not be normalized to 1 at the same time as Lee's solution, no matter how one tweaked the normalizations of the bare states: This was the famous ghost state. They could further show that the transition $V + \theta \rightarrow \text{ghost} + \theta$, i.e., from a regular (properly normalized) to a ghost state, had a non-vanishing amplitude and that consequently probability was not conserved. The Lee Model was not a consistent quantum theory.

To unclutter their discussion, Pauli and Källén had somewhat simplified Lee's model by assuming that the two nucleons had equal mass, i.e., $m_V = m_N$. In that case, the energy of the ghost state (the "mass" of the ghost particle) takes a large negative value proportional to e^{1/g^2}.[4] This expression cannot be expanded in an analytic power series in the coupling constant and the ghost solution would thus not have been found in a perturbative calculation. This guided Pauli's intuition that one might have missed similar ghost states in quantum electrodynamics, where one had no exact, only (the first few terms of) perturbative solutions of the theory (Letter from Pauli to Källén of 4 December 1954, PSC IV-II). It was the simplifying assumption of the mass degeneracy of V and N that Heisenberg proposed to drop, in order to get a version of the Lee Model in which the S-Matrix was unitary after all, despite the persistent presence of the ghost state.

3.1.2 The Battle of Ascona

For his work on the Lee Model, Heisenberg relied on the help of his postdoc Rudolf Haag. Haag had arrived in Göttingen in May 1956, shortly after Nishijima.[5] Haag recalls that he was somewhat more open-minded towards Heisenberg's work than the rest of his colleagues:

> I was essentially of the same opinion as my friends of the "Feldverein" but had some more sympathy with Heisenberg's ideas and was looking forward to discussions with him. [...] A simplified model of a quantum field theory had been devised by T. D. Lee. [...] It could serve as a testing ground for the problems mentioned. Heisenberg asked me to look into it. This was the beginning of a brief but very intense collaboration between us. (Haag 2010, pp. 272–273)

Ultimately, Haag also did not publish with Heisenberg, but they did obtain some novel results, which Heisenberg ended up publishing by himself (Heisenberg 1957a) and which shed new light on Heisenberg's theory. If one dropped the assumption that the two nucleon masses should be equal, one got to tune the mass of the ghost state by modifying the (bare) mass of the V meson. In particular, one could manufacture a situation where the mass of the ghost and of the physical V meson were equal. Heisenberg believed that he had here found an instance of a dipole ghost of the

[4]Pauli consequently referred to this very low-lying state as the "crevasse" (*Gletscherspalte*), Letter from Pauli to Källén of 4 November 1954, PSC IV-II.

[5]Letter from Wolfhart Zimmermann to Kurt Symanzik, 30 October 1955, WZP, Folder Symanzik Correspondence.

sort that he needed in this theory to justify the regularized commutation relations. Heisenberg now hoped to show that for the case of a dipole ghost, the S-Matrix in the Lee Model would turn out to be unitary, lending plausibility to the assumption that this would also be the case in his model, despite the presence of ghosts and the indefinite metric.

This proposal engendered a long discussion between Heisenberg and Pauli. The central question soon turned out to be what happened to the process $V + \theta \rightarrow$ ghost $+ \theta$ in the dipole limit. This was the process that Pauli and Källén had used to demonstrate non-unitarity in the Lee Model with degenerate nucleon masses. In Heisenberg's setup there was no longer a well-defined physical V particle—the V particle and the ghost had merged to form a dipole ghost.[6] Rewriting the eigenvalue (time-independent Schrödinger) equation in this sector as $\chi(E) = 0$, the function χ has a double zero at an energy E that gives the mass of the dipole ghost.

Pauli had originally been directly convinced by Heisenberg's argument that this sector (i.e., the sector with nucleon number 1 and a value of -1 for the second quantum number) came out unitary in the dipole case, and had only been concerned about unitarity (and the possible appearance of further ghosts) in higher sectors (Letter from Pauli to Heisenberg of 28 December 1956, PSC IV-III). But soon discussions with Fierz and Källén had convinced Pauli that even this assertion of Heisenberg's was non-trivial and began to call into question whether Heisenberg's idea would lead to even partial success (Letter from Pauli of 15 January, PSC IV-IV). Heisenberg in his reply of 17 January 1957 now provided for the first time a sketch of the proof of unitarity for the $V\theta$-sector, which he was working out together with Haag. It would become the central focus of the Heisenberg-Pauli debate and should thus be worked out in some detail.

The general form of the states $|\Phi\rangle$ in this sector, which physically describe the scattering of a θ meson on a V nucleon, could again be written down quite straightforwardly, since the only bare particle states with the correct quantum numbers were $|1, 0; 1_k\rangle$ and $|0, 1; 1_{k_1}, 1_{k_2}\rangle$:

$$|\Phi\rangle = \int \varphi(\mathbf{k}) d^3k \, |1, 0; 1_k\rangle + \int\int d^3k_1 d^3k_2 \varphi(\mathbf{k_1}, \mathbf{k_2}) \, |0, 1; 1_{k_1}, 1_{k_2}\rangle \quad (3.13)$$

Plugging this into the Schrödinger equation, one gets an integral equation for $\varphi(\mathbf{k})$, whose general form was given by Heisenberg in his letter:

$$\varphi(\mathbf{k})\chi(\omega) = \int d^3k' \varphi(\mathbf{k'}) f(\omega, \omega') + \text{inhomogeneous term} \quad (3.14)$$

The inhomogeneous term depends on the precise choice of solution for $\varphi(\mathbf{k_1}, \mathbf{k_2})$ and was of no immediate concern. χ is the aforementioned function appearing in the

[6] In a letter of 2 January 1957, Pauli had worried about the loss of V as an actual particle and whether Heisenberg, in his full model, might not accidentally lose the electron. But this was just a side remark and was not directly relevant to the specific question whether the dipole Lee Model was unitary.

eigenvalue equation of the V sector—its zeroes gave the energies of the V particle states. In the usual (Pauli-Källén) case it had two distinct zeroes, corresponding to the masses of the physical V particle and the ghost. In Heisenberg's dipole case, however, χ had a double zero at the energy E_0 of the dipole ghost. Due to the fact that χ has zeroes, φ will be singular in general, a well-known fact in any quantum mechanical scattering problem. The exact type of the singularity cannot be determined from the above integral equation. The usual procedure employed in this case was to fix the exact singularity of the scattering state wave function by imposing the boundary condition that there should be no incoming (only outgoing) spherical waves.

Heisenberg now claimed in his letter that this condition became useless in the case of the dipole ghost. The singularity should rather be determined by the condition that the dipole ghost states do not contribute to the S-Matrix, leaving a matrix that only describes scattering processes involving an N nucleon and two θ mesons, a matrix that involved no negative probabilities and was unitary all by itself. On this procedure, Pauli remarked in the margins of Heisenberg's letter: "But how?"

In evaluating the debate that ensued, there is a strange dynamics at work. Heisenberg was convinced that the program should work, because his philosophical approach would not work otherwise. In an earlier letter, while still discussing the case of higher sectors, Heisenberg had explicitly anticipated an as yet unavailable mathematical result stating that he had to "assume" this specific result following his "general philosophy" (15 January, PSC IV-IV). What he presented to Pauli was, however, sketchy incomplete reasoning, which Pauli had no patience for, precisely because he knew how strongly Heisenberg was convinced of getting the right result. Heisenberg in fact presented an extended (though still incomplete) analysis of the problem already on 21 January 1957, where he argued that his unusual rule for solution choice corresponded to demanding that the incoming ghost states have zero norm and would therefore not upset the unitarity of the S-Matrix. That such a state can be constructed through a suitable linear combination of states with positive and negative norm is intuitively clear and had been acknowledged as a special feature of the dipole limit by Pauli (Letter of 7 January, PSC IV-IV). But by this time, Pauli no longer believed to be in a scientific debate, but rather that he had to relieve Heisenberg of his delusions. In a letter to Källén of 23 January he wrote that he did not believe in the existence of Heisenberg's solution and that the discussion with Heisenberg was "not physics, but special education" (*nicht Physik, sondern Heilpädagogik*).

Heisenberg tried to fight back, asserting that a new prescription for finding the singularities of φ in Eq. 3.14 was unavoidable. Formally imposing the usual "no incoming spherical waves" condition gave a solution that was physically uninterpretable. Because of the double zero of χ this solution would have a double (rather than the usual single) pole at the energy of the dipole and would consequently not contain just outgoing spherical waves (28 January, PSC IV-IV). To which Pauli only replied that, yes, indeed, the entire theory was pathological and finding pathological solutions hardly implied the existence of benevolent ones (29 January, PSC IV-IV). That same day, Heisenberg announced that he would be leaving for the Swiss town of Ascona on Lago Maggiore and would be staying there for a longer while for health reasons. Pauli consequently suggested (31 January, PSC IV-IV) they take a break from their discussion, only to write a lengthy letter in which he stated that all of Heisenberg's "decrees, edicts, and orders" (*Verordnungen, Erlasse, und Befehle*) could never "*conjure up* solutions," (*Lösungen hervorzaubern*) resorting to the following metaphor:

> If, for example, you don't like a girl, you can simply let her go. But it is not effective to order her to be smarter or more beautiful. That will change nothing.[7]

And even though he closed his letter with the words:

> Now I urgently advise you to focus entirely on your convalescence, for which I wish you all the best, and to let this matter rest for the time being. Concerning the latter (letting it rest), *I* will certainly do that.[8]

the tirade that had gone before belied his willingness to let things rest. And Heisenberg was certainly not willing to let things rest either. The sheer number of crossed letters shows the heatedness of the debate, and Heisenberg wrote to Pauli, again on the very same day, providing a sketch of an existence proof for his solution, which he could not yet, however give with "full mathematical rigor." The exchange of letters that followed was later called the "Battle of Ascona" by Heisenberg (1969, p. 306) in his autobiography, describing it as follows:

> My correspondence with Wolfgang from Ascona remains a most painful memory. Both of us fought remorselessly and summoned up all our mathematical resources in an attempt to break the deadlock. At first, my proof was not yet fully clear and Wolfgang could not see what I was getting at. Again and again, I tried to re-present my arguments, and each time Wolfgang was incensed at my failure to see his objections. (Heisenberg and Pomerans (translator) 1971, p. 225)

But this recollection ignores the underlying dynamics that Heisenberg urgently wanted his conjecture to be true, because his program depended on it, and Pauli disbelieved in the conjecture for precisely that reason. In another long letter, completed on 5 February, the first one Pauli sent to Heisenberg in Ascona, Pauli made this abundantly clear:

> Your letter of 31 January finally represents real progress. You admit that you are *not* able to prove the existence of a solution with a simple pole in the Lee Model. (Of course, I don't believe a single syllable of your proof sketch.) In this manner, one can now separate the rational-logical from the irrational (intuitions, expectations, "creeds"). [...]
> I am as good as sure that [...] there will be anomalous states in the final state *after all* [...] For me, this comes from *mathematics*, *not* from a "philosophy"! (An unphysical theory simply has to be treated with great mathematical care and precision).[9]

---

[7] *Wenn Dir z.B. ein Mädchen nicht gefällt, kannst Du sie einfach gehen lassen. Du kannst ihr aber nicht mit irgendeiner Wirksamkeit Befehle erteilen, dass sie schöner oder intelligenter zu werden habe. Es wird sich dadurch gar nichts ändern.*

[8] *Nun rate ich Dir dringend, Dich ganz deiner Erholung zu widmen, zu der ich Dir das Beste wünsche, und die Sache vorerst ruhen zu lassen! Was das letztere betrifft (das Ruhenlassen) so werde ich es ganz bestimmt tun.*

[9] *Dein Brief vom 31. Januar ist wieder ein reeller Fortschritt! Du gibst zu, dass Du die Existenz einer Lösung mit einfachem Pol beim Lee-Modell nicht beweisen kannst. (Von deinem Beweisansatz*

But already at this point, where the debate was at its sharpest, the tide began to turn. Heisenberg's reply from Ascona (6 February) suddenly seems to have clarified for Pauli a point that Heisenberg had already made on 21 January. The initial conditions he was imposing were not supposed to contain no ghost states at all; in fact he was allowing for the presence of incoming spherical waves composed of (ultimately harmless) zero-norm ghost states. Pauli now accepted the existence of Heisenberg's proposed single-pole solution (9 February), but could not see what this existence was good for: Certainly if one had one process for which one added the right amount of fine-tuned zero-norm ghosts, further scattering processes involving the final states of this process would no longer fulfil this condition:

> Then the initial state of the second process will not contain the unphysical $V_0 + \Theta$ states [...] in the right admixture with the physical states in order to obtain unitarity of the S-Matrix for the second process. And there will be no solution to correctly represent the *total* scattering process (the earlier *and* the later one).
> I am therefore at exactly the same point as I was before![10]

After some further back and forth, Heisenberg explained that he was postulating gratuitously adding zero-norm states to each individual scattering process (16 February). And after some further insistence from Heisenberg, expressing his wish that their debate must not end like the debate on quantum mechanics between Einstein and Bohr at Solvay 1927 (28 February), Pauli gradually gave in. On March 1, Pauli, after having tried to end their discussion several times, conceded that their discussion could now continue without the danger of further "resentment" (*Verstimmung*). The immediate follow-up letter was written on two consecutive days (2/3 March) and while on the first day Pauli still characterized Heisenberg's work as "wishful thinking in Mathematik [sic]" and brought forth a last barrage of questions, on the second day found that he could answer most of his questions himself, concluding:

> So now I have learned something: Things are *not* the way you presented them in Pisa, i.e., that an *arbitrary* initial state *without* B [negative norm] states will remain free of B states in the course of time [...]. That was *wrong*! Instead: one needs to tune the initial state by adding (unmeasurable) *in*coming A [zero norm] waves so that the final state [...] contains no B states. That is *not a selection rule at all* [...]; but it appears to have the great advantage of being mathematically correct.[11]

---

glaube ich natürlich keine Silbe.) Damit lässt sich aber nun das Rational-Logische vom Irrationalen (Intuitionen, Erwartungen, "Glaubensbekenntnisse") trennen. [...] Ich bin mir so gut wie sicher, dass [...] im Endzustand doch die anormalen Zustände auftreten [...] Das kommt bei mir aus der Mathematik, nicht aus einer "Philosophie"! (Eine unphysikalische Theorie muss eben mathematisch sehr vorsichtig und exakt behandelt werden). PSC IV-IV.

[10] *Dann werden im Anfangszustand des zweiten Prozesses die unphysikalischen* $V_0 + \Theta$ *-Zustände [...] nicht in der richtigen Beimischung zu den physikalischen Zuständen vorhanden sein, um Unitarität der S-Matrix beim zweiten Prozess hervorzubringen. Und es wird keine Lösung geben, um den* totalen *Streuprozess (den früheren und den späteren) richtig darzustellen.*
Ich bin also wieder genau dort, wo ich vorher war!.

[11] *Nun habe ich also etwas gelernt: es ist nicht so, wie du es in Pisa erzählt hast: dass nämlich aus einem* beliebigen *Anfangszustand ohne B-Zustand auch im Lauf der Zeit [...] kein B-Zustand entsteht. Das war falsch*! *Sondern: man soll den Anfangszustand durch Hinzufügen (unmessbarer)*

But while Heisenberg had convinced him that the Lee-type toy model worked, the question of Heisenberg's own non-linear spinor theory was still wide open:

> After all, the central problem remains *the mathematical existence of the Heisenberg models*. [...] But at the moment I know of no mathematical methods that would convince us *both* and for the time being, we will both have to stay with *our own* respective opinion on this matter.[12]

And even after several further details had been worked out (also in a personal meeting after Heisenberg's stay in Ascona) and Pauli conceded that even Källén was not able to catch Heisenberg in the Lee model (12 April), he continued to emphasize his skepticism concerning the consistency of Heisenberg's full model. Still, for Heisenberg the victory in the Battle of Ascona, a victory of his philosophical approach, was very important and he highlighted it accordingly in his autobiography:

> After nearly six weeks of the most intense effort, I finally succeeded in breaching Wolfgang's defenses. He now realized that, far from trying to produce a general solution of the mathematical problems under discussion, I was merely offering a series of special solutions for which I claimed only that they lent themselves to physical interpretation. We had taken the first step toward a reconciliation, and after working through the various mathematical details, both of us were finally satisfied that the problem had been solved, or rather that the unconventional mathematical schema on which I wanted to base the theory of elementary particles did not contain any obvious self-contradictions. Admittedly, this in itself was no proof that my scheme was a useful one, but there were additional reasons for believing that the solution had to be sought along the lines I had followed, and that I was justified in continuing along them. (Heisenberg and Pomerans (translator) 1971, p. 225)

Still, no-one wanted to co-author the paper on the Lee Model that Heisenberg then completed: Haag was supposedly taking to long to write his section (Heisenberg to Pauli, 5 June 1957, PSC IV-IV), but in his later recollections, Haag stated that he was still skeptical whether the unitarity proof could be extended to higher sectors and thus declined co-signing the paper (Haag 2010). Pauli also declined to co-sign the paper, not because he disbelieved any of the mathematics, but because he did not want to give the impression that he believed there was anything actually physical about Heisenberg's ghost approach (Letter to Källén, 15 July 1957). But he was wavering, and Heisenberg recalls that it was not just himself who had been impressed by the Battle of Ascona: Pauli too was now optimistic that one might construct self-consistent relativistic quantum theories (Heisenberg 1969, p. 314).

It cannot be emphasized enough that nothing had been proven concerning Heisenberg's theory itself. Despite the mathematical sophistication of the Heisenberg-Pauli debate, all that was at stake was a proof of principle that a possible objection to Heisenberg's theory might be circumvented, that there was a theory with ghosts in which those ghosts could be contained. But for Heisenberg, the victory in the Battle

einlaufender A-Wellen so einrichten, dass der Endzustand [...] keinen B-Zustand enthält. Das ist gar keine Auswahlregel [...]; es dürfte aber den großen Vorzug haben, mathematisch richtig zu sein.

[12] *Das zentrale Problem bleibt ja die mathematische Existenz der Heisenberg – Modelle. [...] Aber im Moment weiss ich da keine mathematischen Methoden, die uns beide überzeugen könnten, und vorläufig wird daher jeder bei seiner Meinung darüber bleiben.*

of Ascona was the decisive breakthrough, showing as it did that his philosophical program had won out over the seemingly apparent mathematical impossibility of carrying it through. So complex was the mathematical analysis of quantum field theories that showing the potential refutability of possible arguments against their consistency was already to be considered a success. Recall that Pauli believed that the ghosts in Lee Model might even provide hints toward the inconsistency of QED, by far the most successful quantum field theory of the day. And not just on the side of mathematical consistency was the merit of a quantum field theory hard to assess. Similar difficulties were encountered in assessing its empirical adequacy, as we shall see in the next section.

3.2 Empirical Adequacy

We have so far said very little about the empirical content of Heisenberg's theory, which is of course in keeping with the essentially non-empirical manner in which the theory was constructed. But, of course, establishing the connection of his theory with the empirically observed spectrum of elementary particles and their interactions was a top priority for Heisenberg, which he pursued all through the 1950s.

When viewed through the lens of perturbation theory, the empirical content of Heisenberg's model appeared to be irredeemably inadequate: it described one type of spin 1/2 fermion, which was massless and had a quartic self interaction, giving rise to non-trivial scattering processes between two such fermions. It was, however, one of Heisenberg's central assumptions that perturbation theory would be inadequate to evaluate the empirical predictions of his theory. This assumption was not unfounded: when analyzed perturbatively, the theory gave rise to divergent terms already at first order in perturbation theory, which could not be removed through renormalization methods. If one did not simply accept this fact and dismiss Heisenberg's theory as non-renormalizable, the only other option was to assume, as Heisenberg did, that the divergences were artefacts of perturbation theory and that non-perturbative effects (such as the hypothetical ghost states) would actually render the theory finite. Early on, Heisenberg thus focused on non-perturbative methods for the evaluation of his theory's empirical predictions.

He was not alone in this. The apparent inadequacy of the highly successful perturbative methods of renormalized QED when applied to the nuclear interaction was a central problem of 1950s high energy theory: physicists had found that if one applied perturbation methods to the nuclear interactions and fitted the thus derived values for physical properties to experiment, one obtained nuclear coupling constants far larger than 1, thereby undermining the original perturbative approach.[13]

[13] An early example is the calculation of the anomalous magnetic moments of the nucleons by Borowitz and Kohn (1949): Calculating in exact analogy to Schwinger's calculation of the anomalous magnetic moment of the electron, they found that for their second-order perturbation calcu-

A prominent non-perturbative approach was to split up the involved quantities differently, not according to the power of the coupling constant, but according to the number of particles involved. A quantum state $|\Psi\rangle$ was then characterized by a (infinite) set of "wave functions" with a specific particle number: the wave function of the state for two fermions and one anti-fermion, e.g., would then be given by the matrix element

$$\tau_{\alpha\beta\gamma}(x_1, x_2|y_1) = \langle 0|\, T\psi_\alpha(x_1)\psi_\beta(x_2)\overline{\psi}_\gamma(y_1)|\Psi\rangle \tag{3.15}$$

where $|0\rangle$ is the vacuum state, T the time ordering operator, and the indices are Dirac spinor indices. In Heisenberg's formulation, one then obtained coupled differential equations for these matrix elements from the (Heisenberg) operator equations for the field operators ψ. The wave functions obtained from this coupled set of equations (which would be truncated by setting all wave functions with particle numbers greater than some fixed number to zero) could then be used to calculate the energy of the state, as well as scattering amplitudes for the particles involved. Such methods, going by the name of Tamm-Dancoff approximation (or "new Tamm-Dancoff approximation" for a refined version developed by Dyson 1953) were quite popular tools in nuclear meson physics in the early 1950s, and Heisenberg not only latched on to them; he and his group also provided independent contributions to their development. However, by the mid-1950s, these methods had become rather unpopular within the mainstream high-energy physics community. In the introductory talk of the Sixth Rochester Conference on High Energy Nuclear Physics in 1956, Marvin Goldberger remarked that:

> This work [by Geoffrey Chew in 1953, not by Heisenberg] had the unfortunate effect of lending undeserved support to the Tamm-Dancoff approximation as applied to relativistic field theories. For during the subsequent period of uncertainty about the experimental data, the theoreticians floundered through a morass of old and new, covariant and non-covariant Tamm-Dancoff calculations, none of which added very much to our understanding either of experiments or field theory. (Ballam et al. 1956, p. I-1)

A somewhat more nuanced assessment was given by Francis Low at the International Congress on Theoretical Physics in Seattle several months later:

> I don't want to say anything against this method, because it is an extremely good one for calculating when the coupling constant is weak. When the coupling is strong, however, nobody knows how to calculate anything in relativistic field theories. (Low 1956, p. 220)

The Tamm-Dancoff approximation was thus a refinement of the usual perturbative techniques,[14] but had no insights to offer for theories that were outside the perturbative regime. As physicists in the US, notably Goldberger himself, shifted

---

lations to be valid, they would have to set the coupling constant to 7 (to fit the neutron magnetic moment) or even 52 (to fit the proton).

[14] As is obvious from a representation of the perturbation series in terms of Feynman diagrams, there is a close correlation between the order of perturbation theory and the number of (virtual) particles involved.

towards the novel dispersion methods (Cushing 1990), which promised genuinely non-perturbative insights, Heisenberg stuck to his Tamm-Dancoff guns.

Now, Heisenberg's theory really consisted of two central elements: the Hamiltonian and the postulated regular anti-commutator of the spinor field (Eq. 2.8). The latter should actually have been the *result* of solving the dynamics implied by the Hamiltonian, e.g., through the Tamm-Dancoff approximation. However, since he could not renormalize his theory, Heisenberg needed regular anti-commutation relations from the start and therefore *assumed* the correctness of his anti-commutation relation in his Tamm-Dancoff calculations, creating an intractable mix of approximations and assumptions whose self-consistency he was unable to prove unambiguously (see Heisenberg et al. 1955, p. 446). In this manner, he was able to qualitatively reproduce several features of the known spectrum of elementary particles: the existence of massless vector bosons (photons), a coupling constant for the long-range interactions mediated by these vector bosons that was much smaller than 1 (the fine structure constant α), and massive boson states (pions) that were about a factor of 10 lighter than the lightest fermion states (nucleons). Heisenberg was quite pleased with this, as he wrote to Pauli on 11 May 1955:

> Of course I know that the devil called "agreement with experience" can badly lead one astray; but still I just can't believe that all these surprises are mere coincidences. The impression they made on me were so strong that I got into the mood: "even if the axiomatics cannot be carried through without contradiction, this theory will relate to the correct one in about the same manner as Bohr's theory of atomic structure related to quantum mechanics."[15]

Of course several central features of established particle physics phenomenology are blatantly absent from the above list, in particular the electron. Heisenberg consequently adopted the position that the theory in its current form was a theory of the strong and electromagnetic interactions of hadrons, and that the leptons and the weak interactions would arise through further refinements of the theory, in particular of the assumed interaction term (Ascoli and Heisenberg 1957, p. 187).[16] However, the results obtained from the Tamm-Dancoff calculations were unsuitable to provide clear hints on how to improve the model. The precise connection between the form of the Hamiltonian and the numerical results obtained was obscured by approximations and numerical methods, which were themselves highly malleable, as witnessed by the following anecdote: In a paper with his Italian postdoc Renato Ascoli, Heisenberg managed to obtain a value of 1/250 for the electromagnetic fine structure constant (i.e., off by a factor of about two) (Ascoli and Heisenberg 1957). But Ascoli later

---

[15]*Natürlich weiss ich, dass einen der Teufel mit dem Namen "Übereinstimmung mit der Erfahrung" übel in die Irre führen kann; aber ich kann mir doch nicht mehr denken, dass all diese Überraschungen reiner Zufall sind. Der Eindruck auf mich selbst war so stark, dass ich in die Stimmung kam: "selbst wenn sich die Axiomatik nicht widerspruchsfrei durchführen lässt, wird sich diese Theorie zur richtigen etwa verhalten wie die Bohrsche Theorie des Atombaus zur Quantenmechanik."*

[16]Heisenberg intermittently touched upon the possibility that his theory might ultimately also be able to address gravitational phenomena, e.g., in a letter to Pauli of 10 February 1958. But this was never systematically pursued, a clear indication of the still marginal status of quantum gravity, even within programs to construct a final theory of microscopic physics.

intimated that they had originally obtained "the value 8, and only after Heisenberg had doctored it up, was the value reduced to 1/250" (von Meyenn 2005, p. 781).[17]

Already in 1950, Heisenberg had invoked another possible criterion for specifying the interaction Hamiltonian:

> What remains unsatisfactory is the apparent great arbitrariness in the choice of the [...] $H(x)$, but one can hope that the right functions will be singled out through special properties, such as simplicity or invariance under specific transformations.[18]

The simplicity criterion was never fleshed out by Heisenberg, but the invariance (or symmetry) criterion would become a central tool of theory construction (or elaboration) for him in the late 1950s. Of course, Lorentz invariance had played a central role for Heisenberg from the start, and we have seen how it helped him turn his philosophical notion of reductive monism into an explicit prescription for a quantum Hamiltonian. In the course of the 1950s, the relation between the symmetry properties of field theories and the observed, phenomenological conservation laws and selection rules was established (Borrelli 2015a), and it gradually became clear to Heisenberg that further invariance conditions for his Hamiltonian were not just necessary in order to constrain its exact form, but also to make it physically viable. For the issue of selection rules was intimately connected with Heisenberg's monistic philosophy: If all matter was to be considered as consisting of some universal substance, one should assume that all possible forms this universal substance can take should be transmutable into each other. But while many such transmutations were observed, giving an empirical grounding to Heisenberg's universal substance hypothesis, not all were. There is a finite number of stable particles that do not simply dissolve into radiation or massless (neutrino) matter. There is the electron. Its stability could be explained through charge conservation, and indeed Heisenberg's Fermi Hamiltonian had a symmetry that could be interpreted as giving charge conservation, not the usual gauged $U(1)$ (since there are no fundamental gauge bosons), but at least a global $U(1)$, which could be interpreted as supplying a universal charge for the fundamental spinors, which in turn might give rise to the observed conserved electric charge (without being necessarily identical with it).

But there was another known, stable, massive particle: the proton. Heisenberg was not the only one to view the stability of the proton as a fact in need of explanation, and around 1950 the idea that a conserved charge (nowadays known as baryon number) was responsible for this began to gain in popularity, originating in the work of Wigner (1949). Heisenberg appears to have first engaged with these ideas after reading a paper by Jordan (1952), published in a special issue of the *Zeitschrift für*

[17]It should be remarked that Heisenberg saw this differently. When he sent the paper with Ascoli to Christian Møller in a letter of 30 November 1956 (WHP, Folder 1774), he wrote: "some mistakes appeared in the calculations, which had to be eliminated, and this also led to a slight change in the value of the fine structure constant."

[18]*Unbefriedigend bleibt einstweilen noch die große Willkür, die bei der Wahl der [...] $H(x)$ zu bestehen scheint, aber man kann hoffen, dass die richtigen Funktionen durch besondere Eigenschaften, etwa durch Einfachheit oder durch Invarianz bei bestimmten Transformationen ausgezeichnet sind* (Heisenberg 1950, p. 259).

Naturforschung celebrating Heisenberg's 50th birthday, in which Jordan attempted to implement Wigner's idea through the symmetries of six-dimensional geometry. Heisenberg recognized how central the idea of ensuring the stability of a proton through the symmetry properties of the Hamiltonian could be for his program. In a letter to Jordan of 26 February 1952, Heisenberg called "the stability of the proton the actual miracle in the physics of elementary particles" going on:

> I therefore find it exceptionally interesting that in your paper you give an invariance group from which the corresponding conservation law follows. Since I am sadly a bad mathematician I would incidentally be grateful if you would write down for me a few simple examples of Hamiltonians or Lagrangians that satisfy your invariance requirement, so, e.g., a Lagrangian that only contains one spinor function ψ but contains this wave function not just quadratically but also in a fourth order term or in some even more complicated manner.[19]

Heisenberg was thus asking Jordan to give him an interaction term for his non-linear spinor theory that would also ensure baryon number conservation. Jordan appears not to have delivered, and Heisenberg would merely reiterate the need for such a Hamiltonian in papers of the mid-1950s:

> In summary, the system of elementary particles that arises from [the Hamiltonian of Eq. 2.1] is apparently simpler and thereby less rich than that of the actual elementary particles; for among the actual elementary particles there are at least two laws of the conservation of charge (electrical charge and nucleon number) [...] The wave equation of actual matter will therefore also have to be a little more complicated...[20]

But for several years this remained merely a programmatic demand. Only in late 1957, buoyed by his victory in the Battle of Ascona, Heisenberg presented a new interaction Lagrangian in a letter to Pauli of 30 October 1957. This Lagrangian, while still only describing the quartic interaction of one spinor field, was supposed to also contain the conservation of baryon number:

$$L = A\left[\left(\overline{\psi}\psi\right)^2 - \left(\overline{\psi}\gamma_5\psi\right)^2\right] \tag{3.16}$$

This Lagrangian indeed has *two* $U(1)$ symmetries: multiplication of ψ by $e^{i\alpha}$, which was already a symmetry of the original Lagrangian and which leaves the two summands invariant individually, and multiplication by $e^{i\alpha\gamma_5}$, which mixes the two interaction terms. Now just as with conservation of charge, there was no reason to believe

---

[19] *Ich finde es daher ausserordentlich interessant, dass Sie in Ihrer Arbeit eine Invarianzgruppe angeben, aus der der zugehörige Erhaltungssatz folgt. Da ich leider ein schlechter Mathematiker bin, wäre ich Ihnen übrigens dankbar, wenn Sie mir einmal ein paar einfache Beispiele für Hamilton- oder Lagrange-Funktionen aufschreiben würden, die Ihrer Invarianzforderung genügen, also z.B. eine Lagrange-Funktion, die nur von einer Spinorfunktion ψ abhängt, die diese Wellenfunktion aber nicht nur quadratisch, sondern auch mit einem Glied vierter Ordnung oder in noch komplizierterer Weise enthält.* PJP.

[20] *Im ganzen ist aber das System von Elementarteilchen, das aus [...] entspringt, offenbar einfacher und damit weniger reichhaltig als das der wirklichen Elementarteilchen; denn bei den wirklichen Elementarteilchen gibt es mindestens zwei Erhaltungssätze der Ladung (Elektrische Ladung und Nukleonenzahl) [...] Die Wellengleichung der wirklichen Materie wird also wohl etwas komplizierter sein müssen...* (Heisenberg 1954, p. 302).

that the conservation law following from the new symmetry had anything to do with the physical conservation of baryon number. Heisenberg was merely demonstrating that a Lagrangian including just one spinor field could be sufficiently rich to provide the necessary amount of conservation laws. How these were then related to the conservation laws of the observed elementary particles was an open question, that might ultimately be answered by using sensible approximation techniques.

Pauli was initially underwhelmed by Heisenberg's proposal, so Heisenberg, now apparently on a roll, attempted to include not just the well-established conservation laws of charge and baryon number, but also the approximate conservation laws that had been established in recent particle physics phenomenology and were valid only for certain kinds of interactions. These were:

- isospin, conserved only by the strong interaction
- strangeness, conserved by the strong and electromagnetic interactions

Heisenberg hoped to incorporate strangeness (which could be considered an exact conservation law, as he continued to assume that his model was not yet able to incorporate the weak interactions) in his new Lagrangian as a discrete symmetry,[21] But what to do about the approximate symmetry of isospin? How to incorporate a quantity that was violated by one interaction but violated by another, if in Heisenberg's model all interactions were to arise from one single term in the Hamiltonian?

Heisenberg had contemplated the option of ignoring isospin altogether (Heisenberg 1957b, p. 277), hoping that it would arise from the dynamics of his theory, but ultimately felt that isospin was fundamental enough to warrant its inclusion in the fundamental equations of the theory. It was impossible to simply endow his fundamental spinor with an additional isospin degree of freedom, since this would have implied a fixed correlation between spin and isospin, which was not seen in nature. He thus briefly considered giving up on his monistic principle and introducing a second spinor, whose sole purpose was carrying isospin and then to combine with the fundamental (iso-neutral) spinor to form any kind of spin/isospin combination possible (Ascoli and Heisenberg 1957). But this renouncement of his monistic starting point appeared to become unnecessary when Pauli realized that the symmetry of Heisenberg's new interaction term was even greater than expected.

Pauli had recently renewed his interest in the neutrino in the wake of the discovery of parity violation. In studying the Lagrangian of the free neutrino (i.e., of a massless spin 1/2 particle), he had discovered a novel symmetry transformation that left the Lagrangian invariant, namely:

$$\psi \rightarrow a\psi + b\gamma_5\psi^c \tag{3.17}$$

where ψ^c is the charge-conjugate spinor, and $|a|^2 + |b|^2 = 1$. Pauli (1957) had then investigated what constraints one could impose on the neutrino's (weak four-fermion) interaction term by postulating that this transformation also left the interaction part of the Lagrangian invariant, with ultimately unspectacular results. However, the Turkish

[21] Following an idea by D'Espagnat and Prentki (1955). See also Karpman (1957).

physicist Gürsey (1958) had observed the isomorphism between Pauli's symmetry group and the group of isospin transformations (both of them being isomorphic to SU(2)), and had proposed the physical identification of the two groups in a letter to the editors of Nuovo Cimento, submitted on 5 November 1957. To this end, Gürsey had modified Pauli's original approach, since all the particles involved in isospin transformations, i.e., the hadrons, are, in contrast to the neutrino, massive. When Pauli now received Gürsey's letter before publication and put it together with Heisenberg's proposed Lagrangian, everything appears to have fallen into place: Heisenberg's ur-fermion was massless, so its kinetic term was invariant under Pauli's group. As it turned out, Heisenberg's specific interaction term was also invariant. And with Heisenberg looking for a way to incorporate isospin, Gürsey's identification of the Pauli Group[22] with isospin symmetry appeared as the natural solution, which Pauli pitched to Heisenberg in a letter of 13 December, initiating a brief but intense collaboration.

It should be noted that the physical identification of the Pauli Group with isospin was again merely based on mathematical analogy. Nowhere in his proposal did Pauli make plausible that the transformations encoded in his group were actually related to the physical property of isospin. But such a structural analogy was already an unexpected success, countering, as it did, the expectation that a monistic theory such as Heisenberg's would in principle never be able to muster sufficient mathematical variability to reproduce the contemporary particle zoo. In a second letter, also dated 13 December, Pauli called it a "splendid success" (*ausgezeichneter Erfolg*) that the isospin group came out "nice and smooth."

Now that the potential of reproducing the known conservation laws of particle physics with Heisenberg's interaction term was established, Heisenberg and Pauli began to work out the details of the theory's symmetry structure in an increasingly feverish exchange of letters and phone calls. For despite the unexpected discovery of something resembling isospin symmetry, the symmetry of Heisenberg's interaction term was still actually both too great and too small. It was too great, because isospin was physically not an exactly conserved quantity; it was only the strong nuclear interactions that conserved isospin. And it was too small, because one of the original $U(1)$ groups (multiplications by $e^{i\alpha}$) was now a subgroup of Pauli's isospin group (for $a = e^{i\alpha}$, $b = 0$) and could thus no longer be simply identified with charge or baryon number, since it was now reserved for the isospin component τ_3. But Heisenberg had nifty solutions for both shortcomings, solutions that would constitute the greatest conceptual legacy of his non-linear spinor theory: the concepts of (dynamical) symmetry breaking and the degenerate vacuum.

Concerning the need to violate the isospin symmetry through electromagnetic interactions (at the time Heisenberg was still working under the assumption that his model could not yet accommodate the weak interactions), Heisenberg realized that while Pauli's group was a symmetry of the classical field theory described by his Lagrangian, it was not a symmetry of his commutation relations, which were only invariant under the $U(1)$ subgroup mentioned above. The commutation relations

[22] This should not be confused with the Pauli Group in quantum information theory.

were also not invariant for the $U(1)$ symmetry represented by multiplying the overall spinor by $e^{i\alpha\gamma_5}$, where this novel feature can be demonstrated most easily. Under this transformation, the anticommutation relation goes over into:

$$\left\{\psi_\rho(x), \overline{\psi}_\sigma(x')\right\} \to \left(e^{i\alpha\gamma_5}\right)_{\rho\tau} \left\{\psi_\tau(x), \overline{\psi}_\xi(x')\right\} \left(e^{i\alpha\gamma_5}\right)_{\xi\sigma} \quad (3.18)$$

The anticommutator is thus only invariant if (considered as a Dirac matrix) it anticommutes with γ_5. Looking at Heisenberg's anticommutator, as given by Eq. 2.8, this is only true for the term involving $\not{k}$ but not for the term that is proportional to the Dirac identity matrix. From a modern viewpoint this is easily understood: the chiral $U(1)$ is a symmetry only for massless theories. The Lagrangian indeed contains no mass term, and is thus invariant under the chiral $U(1)$. Heisenberg's anticommutator was, however, based on the assumption that the full interaction would generate massive states, with the scale of the masses given by m. What Heisenberg was thus effectively proposing was what we would now call chiral symmetry breaking with dynamically generated masses. Heisenberg had a somewhat different reading: from the Tamm-Dancoff calculations in his paper with Ascoli he had come to the conclusion that it was the symmetry-breaking term in the anti-commutator that was responsible for the long-range interactions, i.e., electromagnetism. From this, Heisenberg concluded that isospin symmetry was valid for the strong, but not for the electromagnetic interactions, as it should be. Pauli liked this idea, even though it essentially boiled down to claiming that a symmetry present in the Lagrangian was broken by the still unspecified dynamics of the ghost states. Still, attracted by the emerging symmetry properties and the important role his group played in them ("Of course, I am very happy that my transformation is proving to be so fruitful." Letter from Pauli to Heisenberg of 15 December 1957. "The symmetry and group properties of your L[agrangian] are fantastic." Letter of 18 December 1957.), Pauli even communicated the new ideas, including the breaking of isospin symmetry by the commutation relations, to his ever-critical correspondent Källén ("I was impressed"; "I now no longer doubt that the path we have taken is the correct one." Letter to Källén of 16 December 1957).

But the symmetry breaking, which not only broke the isospin symmetry as it should, but also the $e^{i\alpha\gamma_5}$ symmetry, exacerbated the problem of *lacking* symmetry. Here, Heisenberg introduced his most novel idea, the degenerate vacuum. Nowadays, the idea of a degenerate vacuum is mainly associated with symmetry *breaking*; but Heisenberg's initial idea was to use it to *restore* the dynamically broken chiral $U(1)$ symmetry.[23] For this, Heisenberg wrote his regularized commutation relation in a very schematic form (Letter to Pauli of 9 December 1957):

[23] Since Heisenberg had not spelled out the symmetry breaking dynamics, which are only encapsulated in the regularized anti-commutator, it was not clear what role a degenerate vacuum might already play in the symmetry-breaking dynamics. For Heisenberg and Pauli, symmetry breaking and vacuum degeneracy were initially certainly disconnected, as witnessed, e.g., by a remark by Pauli in a letter of 27 January 1958 (PSC IV-IV), in which he was exploring the possibility of doing without a degenerate vacuum: "After all, it isn't the vacuum state itself that is responsible for the reduction of symmetry..."

$$\{\psi_\alpha(x), \psi^*_\beta(x')\} = a_{\alpha\beta}(x - x') + b(x - x')V\gamma^0_{\alpha\beta} \quad (3.19)$$

The essential features of the regularized anticommutator that this schematic form reproduces are the following: One term, a, that commutes with γ_5 and does not disturb chiral symmetry, and a second term that does (as indicated by the presence of γ_0). And then there is the "sign function" V, an operator that was "connected" with the sign of m in the anti-commutator. Now, indeed, the chiral symmetry breaking term is odd in (i.e., changes sign with) m, while the other term is even. The idea was now to find another operator (O in the first iteration) that anti-commuted with V and did not act on the spin indices of ψ (i.e., commuted with all γ matrices). Then multiplication of the field operators by $e^{i\alpha\gamma_5 O}$ left the anti-commutator invariant and, if the commutation relations between O and ψ were suitably chosen, also the Lagrangian, constituting a new $U(1)$ symmetry of the theory.

But what were these operators O and V and what did they act upon? Heisenberg's suggestion was that one understand them through their action on the vacuum state, of which there were now to be two linearly independent ones, $|\Omega_1\rangle$ and $|\Omega_2\rangle$. V would multiply the first by $+1$, the second by -1, while O would switch the two vacua. But what was to be the action of V and O on the other states in Hilbert Space, Pauli asked on 12 December? In answering this question, Pauli and Heisenberg went in very different directions, sowing the seeds for their break several months later. We will discuss the technical details of their two approaches when we discuss that break. But for the moment, in the winter of 1957/58, any disagreement the two might have had was submerged by Pauli's sudden enthusiasm. For a brief period of not even two months (December/January 1957/58), it was Pauli who was the main driving force behind the program, blowing caution to the wind, as repeatedly emphasized in (Rettig 2014). Indeed, it appears that Pauli at this point was almost drunk with symmetry:

> [T]he outlines of a mathematical picture of the structure of the "elementary" particles is gradually becoming visible, and I now hope that the *new year 1958* will bring *total clarification.*[24]

The apparent mathematical beauty of the emerging scheme certainly resonated with Pauli's recently articulated vision of a convergence of mysticism and science in the Pythagorean tradition (as expounded, in particular, in Pauli 1956). He described Heisenberg's idea of a degenerate vacuum almost in terms of a revelation, remarking to his assistant David Speiser (von Meyenn 2005, p. 777): "The vacuum is degenerate; you know: Heisenberg, he perceives [*merkt*] such things!" The doubling of the vacuum resonated with a "mirror complex" of Pauli's, a subconscious and not scientifically motivated preoccupation with mirroring, which he had recently identified

[24] *[D]ie Züge eines mathematischen Strukturbildes der "Elementar"teilchen werden allmählich sichtbar und ich hoffe nun, dass das neue Jahr 1958 die völlige Klärung bringen wird!* Letter from Pauli to Källén, 19 December 1957.

in his correspondence with C. G. Jung.[25] In a letter to Jung's secretary Angela Jaffé of 5 January 1958 Pauli wrote:

> During these days, which were very turbulent for me, I did not have the time to talk on the phone nor to contact you in any other way: a new physico-mathematical theory of the smallest particles is in the process of being created, and I am collaborating closely with Heisenberg on it. ([...] [T]he contribution of a Turk (!) whom I do not know personally and who is currently in Brookhaven was very important for me.) A flood of letters is going back and forth between Göttingen and Zurich, accompanied by a telephone call from Göttingen—about mathematics!
> So, you see, dear Ms Jaffé: "the line is busy".
> There is *a lot* of mirroring going on (literally–as a mathematical operation) [...]
> Heisenberg is a very different person than myself; but we can work together so well because we are *seized by the same archetype*. [...]
> I had been prepared for all this by a *dream* [...]: "In our master bedroom I suddenly discovered two children, a boy and a girl, both of them blonde. They are very similar to each other–as if they had still been the same just a little while earlier–and they both said to me: 'We have been here for 3 days already. We like it here a lot, but no one has noticed us so far." [...]
> This dream greatly agitated me for many days. "The three days" immediately made me think of a dinner I had had with Heisenberg in Zurich exactly 3 days earlier, when he was here just for the layover between two trains. Some ideas of his had impressed me; my "mirror complex" was greatly stimulated by them.[26]

Heisenberg would later recall that he had never seen Pauli like this (Heisenberg 1969, p. 316), and at times even felt that he had to be the moderating force, warning Pauli on 2 January 1958 of "group theoretical castles in the air." Pauli was still in high spirits when he left for the United States on 17 January, where he hoped that others would pick up on their ideas and develop them rapidly, as he wrote to his close friend Paul Rosbaud from the ship:

> Now comes the stage where the principal ideas will be made public. H. and I are expecting this to be followed by a sort of "run". But we feel that it does not really matter *who* then is

[25]Letter to Jung of 5 August 1957. This letter is not reprinted in Pauli's scientific correspondence, but rather in (Meier 1992).

[26]*Ich hatte weder Zeit zu telephonieren, noch sonstwie mit Ihnen in Kontakt zu kommen in diesen für mich stürmischen Tagen: eine neue physikalisch-mathematische Theorie der kleinsten Teilchen ist im Entstehen begriffen, und ich bin in sehr enger Zusammenarbeit mit Heisenberg darüber. ([...] [D]er Beitrag eines mir persönlich unbekannten Türken (!), der zur Zeit in Brookhaven ist, war mir sehr wichtig.) Eine Flut von Briefen zwischen Göttingen und Zürich geht hin und her, noch begleitet von einem telephonischen Anruf aus Göttingen–betreffend Mathematik!*
Also, Sie sehen, liebe Frau Jaffé: "the line is busy". Gespiegelt (wörtlich–als mathematische Operation) wird dabei viel [...]
Heisenberg ist menschlich sehr verschieden von mir; wir können aber deshalb so gut miteinander arbeiten, weil wir vom selben Archetypus ergriffen sind. [...]
Auf all dies war ich vorbereitet durch einen Traum [...]: "In unserem ehelichen Schlafzimmer entdeckte ich plötzlich zwei Kinder, einen Bub und ein Mädchen, beide blond. Sie sind einander sehr ähnlich–so wie wenn sie bis vor kurzem noch ein und dasselbe gewesen wären–und beide sagten zu mir: 'Wir sind schon 3 Tage hier. Wir finden es hier sehr nett, es hat uns nur niemand bemerkt.' [...]
Ich war über diesen Traum sehr aufgeregt, viele Tage. "Bei den drei Tagen" fiel mir sofort ein, dass ich genau 3 Tage vorher mit Heisenberg in Zürich zu Abend gegessen hatte, als er–nur zwischen 2 Zügen–auf der Durchreise hier war. Einige Ideen von ihm hatten mir Eindruck gemacht; mein "Spiegelkomplex" war mächtig angeregt durch sie."

the *first* to develop the further mathematical consequences. We have established our priority regarding the principal ideas, the rest appears to be of secondary interest to us. It is more important that the "boy" develop as fast as possible. Let us put it this way: the boy is now going to school. It is anyway too much for *one* person, probably also too much for two.[27]

Pauli's talk on the non-linear spinor theory in New York on February 1 marked a turning point in the Heisenberg-Pauli collaboration, and we will discuss it in the next section. First, it is, however, important to briefly take stock of what Heisenberg (now joining forces with Pauli) had actually achieved in the years 1957/58: he had been able to show that theories with an indefinite metric, such as his non-linear spinor theory, could not be dismissed simply on grounds of unitarity violation–there were in fact instances of models with an indefinite metric that still supplied (at least as far as one had been able to check) a unitary S-matrix. And he had been able to show that even a model containing just one single spinor field could be endowed with a symmetry structure that appeared to correspond, at least roughly, to the symmetries implied by the conservation laws of contemporary particle physics.

Now this may not seem like much, and the temptation is great to simply dismiss Heisenberg's and Pauli's enthusiasm as delusions of grandeur for the former and number mysticism by the latter, perhaps, in Pauli's case, even tinged by the subconscious realization of impending death.[28] But given the state of quantum field theory at the time, this is an unfair assessment: the mathematical consistency of even the best-established theory of the time, QED, was heavily in doubt at the time, and due to the great mathematical complication the debate on the matter was often delegated to toy models, such as the Lee Model. And given the lack of methods for extracting reliable empirical predictions from QFTs, it was hardly uncommon for contemporary (though in general more phenomenological) theories to rely solely on structural symmetry analogies with observed phenomenological regularities, without working out how those symmetries would actually be reflected dynamically. This should also be kept in mind when evaluating the reception and (to a large extent) the rejection of Heisenberg's theory, as we shall do in the next section.

[27] *Nun kommt das Stadium, wo die Grundideen publik sein werden. H. und ich erwarten eine Art "run" darauf hin. Aber wir finden, dass es nicht viel ausmacht, wer dann zuerst weitere mathematische Konsequenzen entwickelt. Wir sind hinsichtlich der Priorität für die Grundideen nun gedeckt und der Rest scheint uns sekundär. Es ist wichtiger, dass der "Bub" sich möglichst schnell entwickelt. Drücken wir die Sache so aus: der Bub geht nun in die Schule. Es ist ohnehin alles zu viel für einen, aber wahrscheinlich auch zu viel für zwei.* Letter to Rosbaud of 22 January 1958, PRP, Rosbaud-Pauli Correspondence.

[28] This interpretation is given by Konrad Bleuler in his recollections. In a manuscript entitled "Wolfgang Pauli and Werner Heisenberg: personal memories", dated 1985, he writes: "Meeting Pauli many times during this really critical period and realizing the incredible intensity of his efforts, I cannot help thinking that he must have felt - signs of his fatal disease were in fact apparent—but did not really know that his time for work was by now limited." (NBLA).

Chapter 4
Reception and Rejection

Pauli presented Heisenberg's and his new approach to non-linear spinor theory for the first time to a small crowd of physicists in Milan, on his way to board his ocean liner in Genova, on 18 January 1958.[1] But the first presentation to a large audience was given on February 1 at Columbia University. This debacle is the stuff of physicist's legends and various accounts of it are collected in von Meyenn (2005, pp. 871–872). I have also been able to track down some archival material (DBP) concerning this talk in the handwritten notes of Dieter Brill (now emeritus professor at University of Maryland). These largely confirm the established elements of the Pauli debacle: The sudden wavering ("Now-ah-so-ah"), the growing unease in the audience (Brill notes that Lehmann, who was sitting behind him, remarked: I didn't understand that at all—*Das hab ich gar nicht verstanden*), Pauli's questioning of central elements of the proposal ("Heisenberg proposed degenerate vacuum. I'm not certain that this is necessary.") while lashing out at the at those who found their approach mathematically lacking ("I tried to read some of the rigorous papers. It's easy to be rigorous if one doesn't discuss the problem."). The resistance Pauli encountered in New York initiated a (very gradual) change of heart, which we shall discuss in Sect. 4.2, where we will analyze in detail which aspects ultimately led to his strong rejection of non-linear spinor theory. But first we will discuss the rejection that Pauli encountered and more generally the reception of Heisenberg's ideas by the physics community in the spring of 1958.

[1] As recounted in a letter to Viktor Weisskopf of 16 February 1958.

A. S. Blum, *Heisenberg's 1958 Weltformel and the Roots of Post-Empirical Physics*, SpringerBriefs in History of Science and Technology,
https://doi.org/10.1007/978-3-030-20645-1_4

4.1 General Reception

In his autobiography, Heisenberg recalled that he "did not like the idea of this encounter between Wolfgang in his present mood of great exaltation and the sober American pragmatists" (Heisenberg and Pomerans (translator) 1971, p. 234). And this notion that what Pauli encountered was primarily a rejection of their highly speculative and philosophical edifice by the new pragmatic American spirit of physics survives to this day. However, in his immediate report on the talk in a letter to Heisenberg written on the very same day, Pauli recounted that the local Americans had shown a wait and see attitude, and that the main opposition had come from a group of mathematical physicists from the Institute for Advanced Study (IAS) in Princeton, all of them European, two of them German: Freeman Dyson and the former Heisenberg associates Harry Lehmann and Wolfhart Zimmermann.

As we have seen, LSZ had long opposed Heisenberg's theory, had indeed trained themselves in opposing Heisenberg's castles in the air. Heisenberg had written a letter to Zimmermann, on 6 January, giving the basic outlines of his work with Pauli. On the basis of this letter, a meeting was convened at the IAS on 21 January 1958 at 4 pm in preparation for Pauli's talk 11 days later.[2] During the meeting Frank Yang had asked whether the Heisenberg-Pauli theory contained electrodynamics, to which Lehmann had replied that Heisenberg indeed thought it did, citing Heisenberg's derivation of the fine structure constant together with Ascoli. This appears to have turned off Dyson, who remarked that his opinion on this was "unquotable". Indeed, Heisenberg's derivation of the fine structure constant appears to have reminded Dyson of home, that is Cambridge, where the great astronomer Arthur Eddington had turned to constructing speculative theories, relating cosmology with the structure of elementary particles, and trying to explain why the fine structure constant was the inverse of an integer. After his talk, Pauli attributed Dyson's rejection to his "seeing and fearing 'Eddingtonianism' everywhere" (Letter to Heisenberg of 1 February 1958), a view that Dyson later confirmed. So, one objection of these young mathematical physicists was clearly to Heisenberg's shady approximation methods and the undue trust he placed in the numerical predictions derived from them.

Their other central objection lay with the use of the indefinite metric. As opposed to Pauli, they had not been convinced by the Battle of Ascona, not been convinced that Heisenberg's success with the Lee Model had any relevance for a fully relativistic theory. In his letter to Heisenberg, written right after the lecture, Pauli listed the question "What is the metric in Hilbert Space?" as one of the central questions Heisenberg and he would have to address. In a letter of 29 April 1958 to Markus Fierz, Pauli recalled that Lehmann had been very much opposed to the indefinite

[2]Our knowledge of this preparatory meeting is based on handwritten notes taken during the meeting by Dieter Brill, who was then a graduate student at Princeton University (DBP). I want to thank him once more for making them available to me.

metric in New York in February.[3] And in June, Wolfhart Zimmermann wrote to Heisenberg:

> All these discussions have only confirmed my opinion that it is highly probable that there is no field theory with an indefinite metric that also fulfils Lorentz invariance, unitarity of the S-Matrix and macrocausality. [...] [It] appears to be a general empirical statement: Too weak modifications of the principles Lorentz invariance, microcausality, and positive definite metric make the theory worse rather than better.[4]

This notion that one couldn't just tweak one aspect of relativistic quantum field theory (the positive-definiteness of the metric) and hope to leave all of the other desirable features intact appears to have quite pleased Bohr, who famously chimed in, declaring after Pauli's talk that Heisenberg's setup simply wasn't radical enough. However, it needs to be emphasized that all Zimmermann was voicing was an "opinion" and that there was no way to formally prove the impossibility of an indefinite metric; this was merely an "empirical statement". Just as Heisenberg could not show that his theory worked, LSZ were definitely not able to show that it didn't. It certainly didn't fit into the axiomatic framework they were constructing, but it was on the other hand so close to that framework that it did not seem to provide a promising starting point for constructing a consistent novel QFT.

Pauli's own wavering during the legendary Columbia talk obscures the fact that there were in fact some physicists who were enthusiastic about the new Heisenberg-Pauli approach. Gürsey wrote to Pauli after the talk:

> I have also received a letter from three young physicists from the Institute (Princeton), that stronghold of experts. They courageously confess to be interested in the Pauli-Heisenberg theory and ask me for preprints. All this indicates that, in spite of the experts' violent objections, you have attained your principal objective which was to free the unprejudiced physicists from the current dogmas.[5]

Another member of the Columbia audience, Yale physicist Gregory Breit, felt com pelled to write to Heisenberg, inquiring about possible experimental implications of Heisenberg's theory:

> I heard Pauli speak at Columbia at the beginning of February regarding your joint work [...] I am, of course, much impressed by your general philosophy and the new encouragement regarding the theory. I have been wondering as to whether there is any hope of seeing by means of it what modifications in quantum electrodynamics you might expect [...] Is it still to early to attempt such a detailed application?[6]

---

[3]That Lehmann's main objection to Heisenberg's theory was the use of the indefinite metric was also independently confirmed to me in private communication by Bert Schroer, who started his PhD work with Lehmann in 1958.

[4]*Alle diese Diskussion haben mich in meiner Meinung bestärkt, dass es sehr wahrscheinlich keine Feldtheorie mit indefiniter Metrk gibt, die Lorentzinvarianz, Unitarität der S-Matrix und Makrokausalität erfüllt. [...] [Es] scheint offenbar ein allgemeiner Erfahrungssatz zu sein: Zu schwache Änderungen der Prinzipien Lorentzinvarianz, Mikrokausalität und positiv definiter Metrik machen die Theorie eher schlimmer als besser.* Letter from Zimmermann to Heisenberg, 16 June 1958, WZP, Heisenberg Correspondence.

[5]Gürsey to Pauli, 6 February 1958, PSC IV-IV.

[6]Letter from Breit to Heisenberg, 7 March 1958, GBP, Heisenberg Correspondence.

Even more enthusiastic was the support that Heisenberg received from the Soviet Union, which arrived in the form of a letter from Lev Landau, sent just ten days after Pauli's Columbia talk. There is no indication that Landau had heard of Pauli's talk; he had simply finally gotten around to reading Heisenberg's earlier papers and was thrilled by Heisenberg's approach of regularizing the commutation relations:

> We have recently thoroughly studied your papers. They are written in a rather difficult fashion, and up to now I had not brought up the courage for them. The brilliant idea of modifying the Green Function of the primary particles in the manner you suggest made a truly shocking impression. I think it can hardly be doubted that, whatever further difficulties might remain, your idea is the true, unexpected solution of the problem. The photon and the fine structure constant are particularly nice. I find the comparison with experiment absolutely satisfactory, given that it is still only a model.
>
> I want to congratulate you from my heart on your brilliant successes. The old guard does not surrender![7]

Indeed in the late 1950s, there was quite some interest in Heisenberg's non-linear spinor theory in the Soviet Union: Already on 19 October 1957, Heisenberg had sent some reprints of his papers to Dmitrii Ivanenko of Moscow University, apparently in response to a request that is no longer extant. And from Dmitrii Blokhintsev of the Joint Institute for Nuclear Research Heisenberg received another request for preprints on 10 February 1958, stating that the "theoreticians in Dubna are very much interested in them." Heisenberg's folder with USSR correspondence (in which all of these letters are to be found) even contains a reply (dated 10 July 1958) to a (not extant) inquiry concerning the non-linear spinor theory by J. I. Granovsky and A. Yanof of the Institute for Nuclear Physics in Alma-Ata, Kazakhstan.

But in the West, physicists were looking to Pauli for a verdict on what to make of their joint theory. Already in the IAS preparatory meeting in January, Oppenheimer's preliminary verdict on the Heisenberg-Pauli theory had been:

> Pauli sees new ways to introduce representations of which we only have a smell. He[isenberg] is so attached to his earlier work that he tries to interpret it in that. Don't apply Pauli's enthusiasm to H's letters.

And after Pauli's talk, on 7 March, Weisskopf wrote to Pauli from CERN in Geneva:

> In the meantime, Heisenberg has been here. [...] His talk and the following discussion were after all rather impressive. [...] When the group theoretical part was discussed, a number of difficulties were addressed, which Heisenberg did not answer well. [...] I only want to add that the independence of integer and half-integer iso- and regular spins was not explained by

[7] *Wir haben in letzter Zeit Ihre Arbeiten gründlich studiert. Sie sind ziemlich schwierig geschrieben, und ich habe früher nicht den Mut dazu aufbringen können. Die glänzende Idee, die Green'sche Funktion der primären Teilchen in der von Ihnen vorgeschlagenen Weise zu ändern, macht einen wahrhaft erschütternden Eindruck. Ich glaube, dass es kaum noch zu bezweifeln ist, welche Schwierigkeiten auch noch zu beseitigen wären, dass Ihre Idee die wirkliche, unerwartete Lösung des Problems darstellt. Besonders schön ist das Photon und die Feinstrukturkonstante. Ich finde den Vergleich mit dem Experiment vollkommen befriedigend, da es sich ja bis jetzt nur um ein Modell handelte.*

Ich möchte Ihnen mit [sic] Ihren glänzenden Erfolgen herzlichst gratulieren. Die alte Garde kapituliert nicht! Letter from Landau to Heisenberg of 11 February 1958, WHP, Folder 1864.

> Heisenberg at all. He always tried to invoke several vacua, but that is a scam! If one introduces several vacua to get different particles, one is doing just as poorly as by introducing different fields. Where is the unified theory? One is back at the old number game with many fields. But probably there is something we (including Heisenberg!) don't understand. Enlighten us.[8]

And Pauli was well aware that it was now on him to decide the fate of the non-linear spinor theory, especially within the physics community. In fact, Heisenberg's grand claims had already traveled beyond the confines of academia and had reached the general public: a Göttingen colloquium talk held by Heisenberg on 24 February had received extensive news coverage, during which Heisenberg's theory got the moniker *Weltformel* attached to it.[9] Pauli reacted by circulating a now famous cartoon "to declare his independence (among physicists)".[10] The cartoon shows an empty square with the caption: "Comment on Heisenberg's Radio-advertisment: This is to show the world that I can paint like Titian—Only technical details are missing." In a letter to Rosbaud, Pauli further elaborated:

> My desire for "glory" is well satisfied; I just want a field of occupation that interests me (scientifically). For this purpose, H.'s optimism was useful, while certain groups of experts were boring me. In contrast, H.'s need for fame is *insatiable*. What does he want to compensate? Of course, he has inferiority complexes (as you emphasize in your letter). The hypothesis of the American physicists is that his trauma results from the fact that during the war he did *not* realize that one could produce plutonium. Incidentally, his theory of superconductivity also failed. Here [in the USA] I am told everywhere: "Nobody would believe H., if you were not in. [...]" I thus provide him with *credibility*.—Whether rightfully so, we shall see. [...] Up until now, I do not see that the program is impossible, as far as elementary particles + electromagnetism is concerned. The further development of the theoretical approaches will show whether it really works or if essential ideas are missing.[11]

---

[8] *Inzwischen war Heisenberg hier. [...] Sein Vortrag und die darauf folgende Diskussion waren doch recht eindrucksvoll. [...] Als der Gruppenteil diskutiert wurde, kamen allerlei Schwierigkeiten zur Sprache, die Heisenberg nicht gut beantwortete. [...] Ich möchte nur hinzufügen, dass die unabhängigen ganz- und halbzahligen Iso- und Normalspins von Heisenberg gar nicht erklärt wurden. Er redete sich immer auf verschiedene Vakua heraus, aber das ist doch ein Schwindel! Wenn man verschiedene Vakua einführt um verschiedene Teilchen zu kriegen, dann ist man doch genauso schlecht dran, wie man mit verschiedenen Feldern wäre. Wo bleibt dann die unified theory? Dann ist man doch wieder beim alten number game mit vielen Feldern. Aber wahrscheinlich verstehen wir da was nicht (inklusive Heisenberg!) Klären Sie uns auf.*

[9] For overviews of the press coverage of Heisenberg's theory, see von Meyenn (2005, pp. 989–994) and Rettig (2014, Sects. 15.3.4 and 15.3.6.1).

[10] *Um meine Unabhängigkeit (unter Physikern) zu deklarieren*, letter to Paul Rosbaud of 5 March 1958, PRP, Pauli-Rosbaud Correspondence.

[11] *Mein Bedürfnis nach 'glory' ist ja gedeckt, ich will nur ein Betätigungsfeld haben, das mich (wissenschaftlich) interessiert. Hierzu war mir H.s Optimismus günstig, während gewisse Experten-Gruppen mich gelangweilt haben. Dagegen ist H.s Ruhm-Bedürfnis unersättlich. Was will er damit kompensieren? Natürlich sind Minderwertigkeitskomplexe bei ihm vorhanden (wie Sie in Ihrem Brief betonen). Die Hypothese der amerikanischen Physiker ist, sein Trauma rühre daher, dass ihm während des Krieges nicht eingefallen ist, dass man Plutonium herstellen könne. Übrigens ist ihm auch die Theorie der Supraleitung schief gegangen. Überall höre ich hier: "'Nobody would believe H., if you were not in. [...]" Ich helfe ihm also zu Kreditfähigkeit —Ob mit recht, wird sich zeigen. [...] Bis jetzt sehe ich keine Unmöglichkeit des Programmes, was Elementar-teilchen +*

While Pauli had sobered up, he had not yet fully made up his mind about the non-linear spinor theory. It would take another month, during which he remained in the USA, now traveling to the West Coast, to Berkeley, for him to fully disavow himself from Heisenberg's theory, thereby also sealing the fate of that theory for the physics community at large. We will study Pauli's gradual disenchantment and its manifold reasons in the following section.

4.2 Pauli's Turnaround

Initially, Pauli's main source of concern was that he had been "hitched to Heisenberg's Tamm-Dancoff methods" (Letter to Källen of 6 February 1958, PSC IV-IV) of which he had long been critical. Indeed, Heisenberg was still proposing to calculate the energy spectrum of their new Hamiltonian with his version of the Tamm-Dancoff approximation and had set several of his assistants in Göttingen to doing just that. Possibly even more problematic for Pauli was that their entire symmetry breaking pattern was based on Heisenberg's regularized anticommutator, which was still merely an assumption, some attempts at proving its self-consistency with Tamm-Dancoff methods aside. Pauli warned Heisenberg (1 February):

> For you it is probably psychologically important that the new work is a continuation of your previous work. For me, on the other hand, that is a much less important aspect. I just want to have *some* mathematically *well defined* procedure.[12]

Pauli demanded that the fundamental equations of the theory, determining the energy eigenvalues of the Hamiltonian and thus the masses of the observed particles, should at least be defined and written down without making use of the Tamm-Dancoff approximation; and that there should be, at least in principle, a method for deriving, rather than postulating, the anti-commutators. Their entire theory, Pauli feared, had not resolved any of the central issues of Heisenberg's non-linear spinor program (1 February):

> The mathematical foundations are not clarified, and one is still calculating with the old bad methods (are those even methods?).[13]

Pauli had over the course of the preceding weeks, often remarked that he had "no intuition for those terrible Tamm-Dancoff approximations" (Letter from Pauli to

---

Elektromagnetismus betrifft. Ob es wirklich geht oder ob noch wesentliche Ideen fehlen, muss eben die Weiterentwicklung der theoretischen Ansätze zeigen. Letter from Pauli to Rosbaud of 5 March 1958, PRP, Pauli-Rosbaud Correspondence.

[12] *Für dich ist es wohl psychologisch wichtig, dass die neue Arbeit eine Fortsetzung Deiner früheren Arbeiten ist. Für mich dagegen ist das sachlich viel weniger relevant. Ich will nur* irgendein *mathematisch* wohldefiniertes *Verfahren haben.*

[13] *Eine Klärung der mathematischen Grundlage erfolgt aber nicht, es wird nur mit den alten schlechten Methoden (sind das überhaupt Methoden?) weitergerechnet.* Letter to Heisenberg of 1 February 1958, PSC IV-IV.

Markus Fierz of 11 December 1957), and Heisenberg's calculations were indeed not guided by strong intuitions, but rather by brute force, labor-intensive calculations–it is no wonder that the paper that Heisenberg ended up publishing (after Pauli's death) had five authors: Such calculations could only be performed with the postdoc manpower and the computational capacities that a large institute such as Heisenberg's could provide. But ultimately the Tamm-Dancoff problem was not a dealbreaker. In his next letter to Heisenberg, of 10 February, Pauli merely demanded that the Tamm-Dancoff calculations be published in a separate paper, which he would not be co-signing:

> [I]t is my wish that the Tamm-Dancoff calculations (to which I have anyhow not contributed) be sent out (and published) *without my name*. There I want to remain a mere observer! You need to separate them off![14]

Pauli was thus backing down on his strong demands about eliminating Tamm-Dancoff and was now merely asking to quarantine it. This was also because he saw that no-one, especially not the sharpest critics of Heisenberg's program, had anything better to offer with regard to the question of calculating energy eigenvalues in a non-perturbative theory. He made this point most strongly in a letter to Källén of 24 February 1958:

> But, whatever the model may be (be it a spinor model or something else) one needs methods to treat an eigenvalue problem that are more decent than the ones we have now. [...] In my opinion also the Feldverein (LSZ) has utterly failed here. [...] Heisenberg is using the Tamm-Dancoff method for lack of something better.[15]

and similarly in a letter to his assistant Charles Enz of 26 February:

> The experts have the habit of assuring one that they themselves are not able to do it and that the others are not "rigorous". If the Feldverein had produced something decent, Heisenberg would never have ended up using the Tamm-Dancoff methods (in which I personally do *not* believe).[16]

Pauli thus, for a brief period, returned to the project. After all, all of the factors that had originally attracted him to Heisenberg's proposal were still in place; it was just that he now had a more sober outlook on its (well-established) shortcomings. Indeed, Pauli felt that his new, more skeptical, approach might actually be beneficial to the research, writing to Heisenberg on 14 February: "Your optimism and my critical

[14] *[E]s ist mein Wunsch, dass die Tamm-Dancoff-Rechnungen (zu denen ich ja auch nichts beitrug) ohne meinen Namen verschickt werden (und erscheinen). Ich möchte bei diesen bloßer Zuschauer bleiben! Ihr sollt diesen abtrennen!* PSC IV-IV.

[15] *Aber, was immer das Modell sein mag (ob Spinormodell oder etwas anderes) man braucht anständigere Methoden, um ein Eigenwertproblem zu behandeln, als diejenigen, welche jetzt vor liegen. [...] Es ist meine Meinung, dass auch der Feldverein (LSZ) hier ganz versagt hat. [...] Heisenberg wendet da in Ermangelung eines Besseren die Tamm-Dancoff-Methode an.*

[16] *Die Experten pflegen einem da nur zu versichern, dass sie selbst das nicht können und dass die anderen nicht "streng" sind. Hätte da der Feldverein etwas Anständiges produziert, so wäre Heisenberg nie dazu gekommen, die Tamm-Dancoff-Methode anzuwenden (an die ich persönlich nicht glaube).* PSC IV-IV.

attitude is maybe a rather good combination." But over the next weeks, Pauli began to have more substantial misgivings concerning the overall coherency of the theory, as he realized that there was a fundamental divide between Heisenberg and him concerning the role of the degenerate vacuum.

We begin by briefly sketching Pauli's understanding of (or perhaps better: vision for) the degenerate vacuum. It is mainly to be found in his correspondence with Heisenberg, but also shines through in the first draft of a joint Heisenberg-Pauli preprint which Pauli completed on 15 January 1958, right before leaving for the US (reproduced in von Meyenn 2005, pp. 849–861). The starting point was to find an answer to the question what the action of the operators O and V (which we introduced after Eq. 3.19) on Hilbert Space states other than the vacuum should be. Pauli had provided a first answer in a letter of 21 December 1957, during the peak of his enthusiasm. Recall that the operator O switches from one vacuum to the other. Pauli now proposed that this switch was to be understood as the "assignment of the 'bra-' to the 'ket-' (Dirac), as all physical quantities are invariant under this (to be explicitly defined) assignment of the "left" to the "right" Hilbert Space." The vacuum Ω_1 was thus simply the dual of the vacuum Ω_2, and the O operator was defined over the entire Hilbert Space as mapping any Hilbert vector to its dual. Through this interpretation Pauli sought to establish an intimate connection between the degenerate vacuum and the structure of the Hilbert Space with indefinite metric, and ultimately a mechanism to ensure what Heisenberg had shown for (some sectors of) the Lee Model, namely that there would be no transitions between states of positive and negative norm. How exactly this was supposed to happen is rather vague and Pauli himself remarked "That is still dark."[17] The general idea appears to have been that the conservation law one obtained from the symmetry generated by the operator O (physically to be identified with charge or baryon number conservation) should at the same time prevent transitions with negative probabilities:

> The following I have not yet calculated and state it rather as a *conjecture*, which still needs to be confirmed by further calculations. There will indeed be a decomposition into two distinct subspaces and one can probably arrange things so that [...] Q and N (baryon number) provide a "superselection rule" between the subspaces. [...] [I]n each subsystem the metric will be that of Bleuler-Gupta [i.e., the metric of QED, which did involve negative-norm states, which could not, however, propagate to infinity and thus never induced negative-probability transitions][18]

It is not hard to see why Pauli had such high hopes for this conjecture: It would after all have tied together the symmetry structure of the model and the taming of the indefinite metric, thereby establishing an intimate connection between the *consistency* issues and the *empirical adequacy* of the theory. Heisenberg, however,

---

[17] This evaluation, and the quote below, is to be found a long letter written to Heisenberg on Christmas over the course of three days, from 25 to 27 December 1957, PSC IV-IV.

[18] *Das folgende habe ich noch nicht gerechnet und sage es mehr als Vermutung, die noch durch weitere Rechnungen bestätigt werden muss. Es wird in der Tat ein Zerfall in zwei getrennte Termsysteme herauskommen, und zwar kann man es wohl so einrichten, dass [...] Q und N (Baryonenzahl) für eine 'Superselection-rule' zwischen den Termen sorgen. [...] in jedem Teilsystem die Metrik die Bleuler-Guptasche ist...*

never actually endorsed these ideas, and his initial response (on 2 January 1958) was hesitant at best:

> You are completely right that the doubling of the vacuum implies that the question of the Hilbert Space metric appears in a different light. However—as we already discussed on the phone—one must *not* connect this metric with Q and N [charge and baryon number] [...] Rather, the metric is only connected with the notion of probability.[19]

But Pauli was not yet deterred. To Källén he wrote on 10 January: "The conservation laws of Q and N ensure that [...] unitary S Matrices exist." It was only when re-evaluating the theory soberly after the shock of the New York talk that Pauli realized that Heisenberg had been pursuing a diametrically opposed path. Not only was Heisenberg not convinced that one could combine Hilbert Space structure and degenerate vacuum; he no longer believed even in the connection between the degenerate vacuum and the conservation laws. Indeed, already in a letter of 30 December 1957, which Pauli had essentially ignored, Heisenberg had distanced himself from his original idea of using the degenerate vacuum to restore the $U(1)$ symmetry that was broken by the anti-commutator:

> The quantity V in the commutation relations is after all at first something like, shall we say, an external electric field **E** in the Hamiltonian of hydrogen. V disturbs the invariance with respect to the [Pauli-Gürsey and chiral $U(1)$ groups], just like **E** disturbs the invariance under spatial rotations. Of course one can perform the corresponding transformation of **E**, i.e., also rotate **E**, then everything remains invariant. But *this* kind of invariance doesn't guarantee any conservation laws [...]. In our case, one can of course also 'rotate' the V [...], i.e., perform the [...] transformation of the vacuum. But that also does not guarantee any conservation laws [...] So, for the time being, I don't see the sources for the two strict conservation laws for Q and N at all.[20]

But why then hold on to the idea of a degenerate vacuum at all? In the same letter in which Heisenberg disavowed himself from the idea of restoring symmetry through the degenerate vacuum, he had identified a new problem for the degenerate vacuum to resolve: the theory was lacking not just in symmetry structures, but also in representation content. This needs some unpacking. In quantum mechanics, one would determine the representation content of a theory (by which I mean the representations under which the states in the theory's Hilbert Space transform under the

---

[19] *Du hast völlig recht damit, dass durch die Verdopplung des Vakuums die Frage nach der Metrik im Hilbertraum in einer neuen Weise gestellt wird. Aber—darüber sprachen wir ja schon am Telefon—man darf diese Metrik* <u>*nicht*</u> *in Verbindung bringen mit* Q *und* N *[...] Vielmehr hat die Metrik allein mit dem Wahrscheinlichkeitsbegriff zu tun.* Letter from Heisenberg to Pauli of 2 January 1958, PSC IV-IV.

[20] *Die Größe V in den Vertauschungs-Relationen ist doch zunächst so etwas wie, sagen wir, ein äußeres elektrisches Feld* **E** *in der Hamiltonfunktion des Wasserstoffs. V stört die Invarianz gegen die Gruppen (A) und (B) ähnlich wie* **E** *die Invarianz gegen Raumdrehungen stört. Natürlich kann man eine entsprechende Transformation von* **E** *vornehmen, d.h.* **E** *mitdrehen, dann bleibt wieder alles invariant. Aber* <u>*diese*</u> *Art der Invarianz garantiert ja keine Erhaltungssätze [...]. In unserem Fall kann man natürlich auch V [...] 'mitdrehen', d.h. die [...] Transformationen am Vakuum ausführen. Aber auch das garantiert keine Erhaltungssätze [...] Ich sehe also zunächst überhaupt noch nicht die Quellen für die beiden strengen Erhaltungssätze für Q und N.*

relevant symmetry transformations), by finding the eigenbasis of some maximal set of commuting operators, one of them being the Hamiltonian. In perturbative QFT, on the other hand, where this was not an option, the representation content was instead determined (e.g., by LSZ) using the asymptotic condition: Asymptotically, any state approached a state composed of a number of free particles. Thus, a complete set of states (and their transformation properties) could be determined by taking all of the free field operators (or the corresponding creation operators) of the theory and acting with them on the vacuum state, to construct first one-, then two-, then general n-particle states.

Now even this construction method for the theory's Hilbert Space was really not available to Heisenberg: In his theory all of the observed particles were actually strongly bound states, while the elementary ψ particles were not supposed to appear in asymptotic states at all. Still, Heisenberg expected to construct his Hilbert Space (and thus determine the representation content) in almost full analogy to weak-coupling QFT with an asymptotic condition: States were to be constructed by acting on the vacuum with (a power series in) $\psi^\dagger$ and ψ, the only difference to QED being that one did not renormalize in order to make $\psi^\dagger |0\rangle$ into a one-particle state. Rather, even the one-particle states were obtained by acting on the vacuum with complicated polynomials in ψ. Combining this view of the Hilbert Space with the theory Pauli and he had constructed, Heisenberg noticed that there was an issue: by identifying the Pauli-Gürsey group (which acted on the spin degrees of freedom) with the isospin, one was inextricably linking spin with isospin, implying that there could be no particles with spin but without isospin or vice versa, in blatant contradiction to the empirically established spectrum of hadrons, which contained both a spin 1/2 particle without isospin (the Λ baryon), and scalar particles with isospin (such as the K mesons). What was even worse, once one had interpreted the state whose leading term was $\psi^\dagger |0\rangle$ as the proton (the corresponding anti-particle being the anti-neutron, due to the identification of Pauli-Gürsey and isospin) it was fundamentally unclear which state to identify with the neutron. The degenerate vacuum offered a whole host of new possibilities here; in particular one could simply identify $\psi^\dagger |\Omega_1\rangle$ with the proton and $\psi^\dagger |\Omega_2\rangle$ with the neutron, thereby effectively identifying the vacuum as an isospin doublet. This further implied that a state could get its isospin either from the field operators or from the vacuum those operators acted on, thereby eliminating the overly restrictive spin-isospin correlation.

This novel reading of vacuum degeneracy was included in the final version of the Heisenberg-Pauli preprint, which was prepared by Heisenberg, based on Pauli's draft and completed on 10 February 1958, a week after Pauli's New York talk and reproduced in (Blum et al. 1993). Heisenberg in Göttingen was now setting his team to work: Heinrich Mitter and Siegfried Schlieder were performing Tamm-Dancoff calculations, while Heisenberg's new assistant Hans-Peter Dürr was "critically analyzing the doubling of the vacuum." In the meantime, Pauli was studying Heisenberg's new draft and the novel use of the degenerate vacuum. By late March, he had convinced himself that the degenerate vacuum was now merely an ad-hoc trick to (at least potentially) get a rich enough particle spectrum. On 20 March, he wrote to Heisenberg:

> The root of the problem appears to me to be *this*. The idea, which we already discussed during the dinner in Zurich, namely: a *general* bisection of the world [...]—that this *more general* idea, in close connection with the division of the Hilbert Space into I and II could hardly be developed: It was in connection with this general idea that I then situated also the degeneracy of the vacuum, which is now just appended as a *trick* to obtain the combination of half-integer isospin (regular spin) with integer regular spin (isospin).[21]

And in a letter to his assistant Charles Enz, who had remained in Zurich, Pauli wrote on 22 March:

> The entire spinor model of Heisenberg and myself appears to me to be losing its attractiveness as I get to know it better (the indefinite metric is not integrated as nicely as I had hoped), but we will probably have to publish something after all, so that nobody has to repeat the same considerations.[22]

To make matters worse, Pauli soon realized that the degeneracy of the vacuum didn't even work as a formal trick, because it could not properly reproduce two-particle states. Pauli outlined his misgivings in a letter to Heisenberg's new assistant Dürr of 26 March 1958. The problem in a nutshell is the following: If indeed $\psi^\dagger\,|\Omega_1\rangle$ is the proton and $\psi^\dagger\,|\Omega_2\rangle$ is the neutron, then $\psi^\dagger\psi^\dagger\,|\Omega_1\rangle$ should be two protons and $\psi^\dagger\psi^\dagger\,|\Omega_2\rangle$ should be two neutrons. But there is no state that can be identified with, say, deuterium, i.e., a state with one proton and one neutron.

A few days after the letter to Dürr (and before a response had arrived), Pauli travelled from Berkeley to Caltech for a few days, where he discussed the matter with Feynman and Gell-Mann. The discussion convinced him[23] that even in the Lee Model the indefinite metric would (as Pauli himself had believed before the Battle of Ascona) lead to difficulties in the higher (quantum number) sectors, which Heisenberg had not studied. This brought Pauli back to the starting point of his involvement with Heisenberg's theory, calling into question the motivation that had originally caused Pauli to soften to it. This was the straw that broke the camel's back, and upon his return to Berkeley (7 April 1958) Pauli wrote a letter to Heisenberg announcing that he no longer wished to pursue their plan of publishing a joint paper on the matter, giving as reasons the three main points that we have discussed:

- He no longer believed that Heisenberg's proposal could be *empirically adequate*, because the degenerate vacuum was an insufficient tool for reproducing the known

---

[21] *Die Wurzel des Übels scheint mir die zu sein, dass die Idee, die wir bereits bei jenem Abendessen in Zürich besprochen haben, nämlich: eine allgemeinere Zweiteilung der Welt [...]—dass diese allgemeinere Idee, in enger Verbindung mit der Teilung des Hilbertraumes in I und II, sich so wenig entwickeln liess: In Verbindung mit dieser allgemeinen Idee habe ich mir damals auch die Entartung des Vakuums gedacht, die jetzt nur als* Trick *angehängt ist, um das Zusammengehen von halbzahligem Isospin (gewöhnlichem Spin) mit ganzzahligem gewöhnlichen Spin (Isospin) herauszubekommen.*

[22] *Das ganze Spinormodell von Heisenberg und mir scheint mir bei näherer Bekanntschaft stark an Reiz zu verlieren (die indefinite Metrik ist nicht so schön eingearbeitet, wie ich gehofft habe), aber wir werden wohl doch etwas publizieren müssen, damit nicht andere noch einmal dieselben Überlegungen wiederholen müssen.*

[23] Letter to Markus Fierz of 6 April, PSC IV-IV.

particle spectrum from a single underlying field. There was "a discrepancy between the mathematical possibilities [...] and the physical facts."

- He disagreed with Heisenberg's (Tamm-Dancoff) methods for *extracting quantitative predictions* from the theory.
- He no longer believed that Heisenberg's treatment of the indefinite metric was *mathematically consistent*, and that Heisenberg had, after winning the Battle of Ascona, decided to rest on his laurels and to no longer think about the indefinite metric at all.

Now, it should be emphasized that Pauli did not wish to see his break with Heisenberg as an endorsement of the LSZ opposition, as he wrote to Fierz on 6 April, while still working on the break-up letter to Heisenberg:

> But let me first make some supplementary negative remarks about LSZ [...] I never thought of holding it against them that they do not join Heisenberg and his ideas. (To the contrary, I always very much liked this strength of character in LSZ, and it was even one of the reasons for me to so strongly recommend Lehmann for the professorship in Hamburg.) But *what I do hold against them is that they do not have any initiative or path of their own*[24]

His invectives against LSZ and the QFT "experts" in general were actually far more severe and elaborate than those against Heisenberg. Consider, the following assessment of Symanzik, the only member of LSZ who was still in Göttingen, which Pauli gave in a letter to Bruno Touschek on April 14 (PSC IV-IV):

> [H]e appears to have reached a position where *"rigor" becomes identical with intellectual nihilism!* Well, since no (physically relevant) statements on quantized field theories can be proven or disproven, it becomes permissible to make arbitrary claims. It is therefore easy for him to live in "rigor" next to Heisenberg in the same institute![25]

On 29 April, in a letter to Enz (PSC IV-IV), he dismissed the entire axiomatic approach[26] to QFT: "[A]xioms are at best useful in a finished theory (and even then their value is questionable); but never do they point to a good way towards further development." An assessment that was supplemented by a couplet, which I feel unable to translate and will leave to the German-speaking reader to enjoy:

> Süß wie Honig entströmt dem Mund der Experten der Humbug.
> Fern vom Grund der Physik scheint er als bläulicher Dunst.

---

[24] *Vorerst aber noch ergänzende negative Bemerkungen über LSZ [...] Nie habe ich daran gedacht, diesen vorzuwerfen, dass sie sich Heisenberg und seinen Ideen nicht anschliessen. (Im Gegenteil hat mir diese Charakterfestigkeit bei LSZ stets gut gefallen, und sie war sogar mit ein Anlass, dass ich Lehmann so sehr für die Professur in Hamburg empfohlen habe.) Dagegen werfe ich ihnen vor, dass sie keine eigene Initiative bzw. keinen eigenen Weg haben.* Letter to Fierz of 6 April 1958, PSC IV-IV.

[25] *...er mir einen Standpunkt erreicht zu haben scheint, bei dem "Strenge" mit geistigem Nihilismus identisch wird! Nun ja, da keine (für die Physik relevanten) Behauptungen in quantisierten Feldtheorien bewiesen oder widerlegt werden können, ist es eben erlaubt, darüber beliebige Behauptungen aufzustellen. Leicht ist es ihm daher in "Strenge" neben Heisenberg im gleichen Institut zu leben.*

[26] "Opportunistic Axiomatics" as Stoelzner (2001) has called it.

Also, Pauli was still interested in central elements of the framework he had developed with Heisenberg, such as the indefinite metric, the identification of the Pauli-Gürsey group with isospin, or the regularized commutators. He merely believed that Heisenberg's radical monism had failed and that the only way to incorporate these features in a model was to introduce more than one field (and ditch the degenerate vacuum): "Farewell unified quantum field theory!", as he wrote to Fierz.[27]

All of this indicates that Pauli was well aware that the problems went far beyond Heisenberg, and that the real issue was the overall state of theoretical physics and QFT in particular, which he describe in to Thirring (letter of 20 May 1958) as "a jungle, which no-one has really been able to penetrate so far."[28] Heisenberg and LSZ were equally misguided in what they were doing:

> [T]he theoretical physicists of that eternally divided people [the Germans] are divided into two halves: one (Heisenberg) has betrayed mathematics, the other (LSZ and followers [*Mitläufer*]) has betrayed physics.[29]

This blink-and-you-miss-it Nazi reference[30] to LSZ is, however, to be contrasted with the crass wording Pauli used when talking about Heisenberg in a letter to Paul Rosbaud of 18 April:

> You have characterized Heisenberg well with "...trying to force things that can't be forced." "And if thou'rt unwilling, then force I'll employ!" is also very German. Göring once called this "ice-cold realism." Ice-cold it was indeed, but the Germans tend to have difficulties with the "realism." Is it realism to want to conquer the entire world, without seeing how one should actually be able to do that as so small a country? Or a different formulation, which I recently wrote to Weisskopf—in connection with H.—: "German endeavors have a marked tendency to end in a Twilight of the Gods, with the Rhinegold sinking back into the river!"[31]

---

[27] *Adieu, einheitliche Quantenfeldtheorie!* Letter of 6 April, PSC IV-IV.

[28] It should be noted that, at least in private, LSZ were willing to admit to the dismal state of and the dismal prospects for axiomatic QFT. In an interview conducted by Dieter Hoffmann and Ingo Peschel [Max Planck Institute for the History of Science, Preprint 485], Peter Fulde, who was a student at the University of Hamburg in the late 1950s, remembers expressing his interest in doing a PhD thesis in field theory and being rebuffed by Lehmann: "Field Theory? They're stuck. Axiomatic field theory is stuck. The best mathematicians are trying their hardest, but no progress is being made." [*Die besten Mathematiker beissen sich die Zähne aus, aber es geht nicht weiter.*]

[29] Letter to Fierz of 6 April; *[...] die theoretische Physiker jenes stets gespaltenen Volkes in zwei Hälften zerfallen, von denen die eine (Heisenberg) die Mathematik, die andere (LSZ und Mitläufer) die Physik verraten haben.*

[30] In postwar denazification, *Mitläufer* was the term used for people who were not directly tied to war crimes, but still had sufficient ties to the Nazi regime not to be exonerated.

[31] *Heisenberg haben Sie gut charakterisiert mit dem "...versuchen, Sachen zu erzwingen, die man nicht erzwingen kann" "Und gehst [sic] du nicht willig so brauch' ich Gewalt!" ist auch sehr deutsch. Göring nannte das einmal "eiskalten Realismus". Eiskalt war es schon, aber mit dem "Realismus" pflegt es dann zu hapern bei den Deutschen. Ist es Realismus, die ganze Welt erobern zu wollen, ohne zu sehen, wie man das eigentlich als ein so kleines Land machen könnte? Oder eine andere Formulierung, die ich kürzlich—im Zusammenhang mit H.—an Weisskopf schrieb: "deutsche Unternehmungen haben eine ausgesprochene Tendenz mit einer Götterdämmerung zu enden, wobei dann das Rheingold wieder im Fluss versinkt!".*

Even if Pauli believed that Heisenberg's research program might still actually be more fruitful than that of axiomatic QFT, at least LSZ were not trying to co-opt him. While motivated by clear intellectual differences, the forcefulness of Pauli's break with Heisenberg was certainly driven by his desire to distance himself from Heisenberg, who was scientifically and politically compromised, in the perception of both the physics community and the general public. So, on the day after breaking with Heisenberg (8 April), he wrote a short memo, which he sent out to a large number of physicists, announcing that the publication was off and giving reasons, though merely emphasizing the first point above (the impossibility of reproducing the elementary particle spectrum with just one field), which was the central point in which Pauli went beyond the criticism he had received from the "experts" in New York.

Similarly, Pauli also tried to distance himself from Heisenberg in the eyes of the general public. The first press reports on Heisenberg's *Weltformel* had been based on a Göttingen colloquium talk for an audience of experts, and Heisenberg had insisted that it had not been his intention to get the press involved. Pauli's reaction to the hullabaloo, which also led some journalists to contact him in the US, was to give curt answers and dampen the enthusiasm (Rettig 2014, p. 170). Heisenberg's reaction was different: he was not so much angered at the media interest per se, but rather by the fact that they were getting it wrong, publishing "dreadful nonsense" (*haarsträubenden Unsinn*).[32] And he knew the perfect venue to set the record straight.

On 25 April 1958 there was going to be a major celebration of the Physical Societies of East and West Germany in Berlin, to commemorate the centennial of Max Planck's birth, an event primarily organized by Max von Laue, Planck's most famous Ph.D. student. And Heisenberg was to be one of the three main speakers, alongside Gustav Hertz (representing East Germany) and Wilhelm Westphal (representing West Berlin).[33] On 4 March, he informed a journalist of the German newspaper *Die Welt* that he would "elaborate his new, mathematically formulated theory for explaining the worldview of modern physics in detail and for a general public in Berlin on 25 April."[34] A few days later Heisenberg retreated on an extended vacation to the island of Ischia, off the coast of Naples (Letter from Heisenberg to Pauli, 13 March 1958). Von Laue only heard through the press that Heisenberg was planning to use the Planck centennial to present his new theory; unable to reach Heisenberg, von Laue contacted Pauli, asking him to intervene, so that Heisenberg would not make the Planck festivities all about himself.[35] But to Pauli the main issue was that his

[32]Letter to Pauli of 5 March 1958, PSC IV-IV.

[33]On the historical background of this event, in particular in the context of German separation, see Hoffmann (1996). I would like to thank Dieter Hoffmann for making the official program of the Planck centennial available to me.

[34]*seine neue mathematisch formulierte Theorie zur Erklärung des modernen physikalischen Weltbildes am 25. April in Berlin ausführlich und allgemein verständlich erläutern.* This article from the *Welt* is reproduced in Eckert (2000).

[35]Letter from von Laue to Pauli of 18 March 1958, PSC IV-IV.

name not be mentioned. In his breakup letter, even before going into his scientific disagreements with Heisenberg, Pauli insisted:

> It will be easy for you to mention my contribution in a footnote of a paper authored solely by yourself. What is much more important to me is *that my name is not written above a thing that I can no longer take responsibility for, and I want to urgently ask you to also take this into account in your lecture at the Planck celebration.*[36]

From the published version of the talk (Heisenberg 1958), it appears that Heisenberg indeed did not mention Pauli as a collaborator, only as a source. But he most certainly did not comply with von Laue's request; indeed, the Planck centennial is now mainly remembered for Heisenberg's presentation of his theory. In any case, Heisenberg's attempts at placating Pauli were unsuccessful, as were Dürr's attempts at repairing the defects that Pauli had pointed out: Dürr first tried to pitch an infinitely (rather than doubly) degenerate vacuum (letter to Pauli of 7 April 1958), but in the paper ultimately published by Heisenberg, Dürr, Mitter, Schlieder and Kazuo Yamazaki, a Japanese Alexander von Humboldt scholar, Pauli's point was conceded (Dürr et al. 1959, p. 443):

> One will thus need at least one more continuous one-parameter group [...] in the mentioned sketch of Pauli and one of the authors it was attempted to get these groups through a doubling of the vectors in Hilbert Space, in particular the vacuum. [...] [T]his no longer appears to be possible to us.[37]

Pauli publicly announced his break with Heisenberg at the 8th Rochester Conference, the first one in the series to be held in Europe, in Geneva, from 30 June to 5 July 1958. This was the first time that Heisenberg and Pauli met after the breakup and Pauli's return to Europe. Pauli, who had chaired the session in which Heisenberg presented, described the events in a letter to Fierz on 9 July:

> [T]he session I chaired was as *satisfactory* as could be expected, and I *rather enjoyed* it. My only concerns were to
> (a) prevent that Heisenberg keeps on going around telling everybody that I agree with him (b) move the attention and the interest away from the spinor model to the more general problem of the indefinite metric [...] Both goals were fully achieved (for the future I now plan to no longer take part in the discussions after Heisenberg's talks. [...])[38]

---

[36]*Du kannst sehr leicht meinen Anteil in einer Arbeit von Dir allein in einer Fußnote anmerken. Viel wichtiger als dies ist mir, dass mein Name nicht über einer Sache steht, die ich nicht mehr verantworten kann, und ich möchte Dich dringend bitten, dies auch in Deinem Vortrag bei der Planckfeier zu berücksichtigen.*

[37]*Man wird daher noch mindestens eine kontinuierliche einparametrige Gruppe [...] brauchen [...] In dem erwähnten Entwurf von Pauli und einem der Verfasser war versucht worden, diese Gruppen durch eine Verdopplung der Vektoren im Hilbert-Raum, insbesondere des Vakuums, zu gewinnen. Dies erscheint uns aber [...] nicht mehr möglich.* The authors did, however, realize that the symmetry breaking scheme, which was still being upheld, in itself already implied a degeneracy of the vacuum, and the concept of breaking symmetries with a degenerate vacuum, inspired by Heisenberg, would go on to have a glorious career in particle physics (Borrelli 2015b).

[38]*war mir die von mir präsidierte Sitzung soweit* <u>befriedigend</u>*, als dies zu erwarten war, und ich fühlte mich* <u>recht wohl</u> *dabei. Es handelte sich nur darum, (a) zu verhindern, dass Heisenberg weiter herumläuft und allen Leuten erzählt, ich sei mit ihm einverstanden, (b) die Aufmerksamkeit*

When Pauli and Heisenberg met again shortly afterwards at the summer school in Varenna, 21 July to 9 August 1958, Heisenberg felt that Pauli had again softened toward their theory, an impression he even stated in his autobiography (Heisenberg 1969, p. 319). But Pauli was insistent that this was not the case.[39] And it is safe to say that while Pauli's break with Heisenberg may not have made an overly strong impression on the general public (with Pauli's name hardly being mentioned in the press anyway), it certainly defined the physics community's view of Heisenberg's theory to this day, where the non-linear spinor theory, if it is known at all, is merely an amusing, if lamentable anecdote. To quote one final letter, to Paul Rosbaud on 3 May 1958, where Pauli wrote:

> So it appears to me that the whole unpleasant journalism will fall back badly on H. For in his case it is *not* relevant what the average newspaper reader thinks, but rather what the physicists think. I have the impression, H. is entirely isolated even among the German physicists (not to speak of the rest of the world; basically *all* physicists know that I will *not* be publishing with him) (also Landau!).[40]

This universal rejection of Heisenberg's theory was given further emphasis by the story's final tragic twist. On 21 November 1958, Pauli was in Hamburg to receive an honorary doctorate. Bert Schroer,[41] then a young PhD student of Harry Lehmann, remembers how Pauli sank into a swivel chair (known as the Pauli armchair, as it had already been at Hamburg University back when Pauli was a professor there), visibly exhausted, exclaiming: "I am having a hard time digesting that Heisenberg." (*Der Heisenberg liegt mir schwer im Magen*). Five weeks later, Pauli was dead of pancreatic cancer. Killing Pauli was the final sin of Heisenberg's *Weltformel*.

---

und das Interesse vom Spinormodell weg auf das allgemeinere Problem der indefiniten Metrik [...] zu lenken. Beide Zwecke wurden voll erreicht (in Zukunft beabsichtige ich nun nicht mehr, in Diskussionen nach Heisenbergs Vorträgen zu sprechen. [...]).

[39] See the discussion of the Varenna meeting in von Meyenn (2005, pp. 1238–1239).

[40] *So scheint es mir, dass die ganze üble Publizistik schließlich schlimm auf H. zurückfallen wird. Denn in seinem Fall kommt es* <u>nicht</u> *darauf an, was der Durchschnitts-Zeitungsleser meint, sondern was die Physiker meinen. Ich habe den Eindruck, H. ist auch in Deutschland unter den Physikern ganz isoliert (von der übrigen Welt ganz zu schweigen; dass ich* nicht *mit ihm publiziere, wissen praktisch* <u>alle</u> *Physiker)(auch Landau!).*

[41] eMail from Bert Schroer to the author of 22 February 2018.

Chapter 5
Conclusions

After having charted the rise and fall of Heisenberg's non-linear spinor theory, it is now time to analyze this story for what it has to say on issues of non-empirical theory construction and non-standard theory assessment, as they are also currently being debated. In the following, I will be looking more closely at four elements of the Heisenberg story. In all four cases the analysis will follow a similar pattern, so it makes sense to begin by outlining the pattern: I identify a central element (A) of Heisenberg's decision to pursue the non-linear spinor theory and of his belief that he was on the right track in doing so. I then briefly reconstruct how the point in question can be traced to a certain feature (B) of research in quantum field theory and particle physics in general, as detailed in this book's central narrative.

Having thus established a causal connection between a feature of QFT research in general (B) and a feature of Heisenberg's specific research program (A), I then draw a historical analogy, arguing that the general feature (B) of research in QFT is still with us today and that we also still observe research programs being pursued that share the element in question (A) with Heisenberg's program. These examples will primarily be drawn from string theory, but that is only because of that theory's prominence and the resulting availability of (at least some) historiographical reflection. In any case, for the contemporary analogue I cannot directly establish the causal connection between (A) and (B) that I established for the Heisenberg story; not just because I have not analyzed the contemporary analogue historically, but also because that causal connection will probably be somewhat obscured in more elaborate and mature research programs. The general argument will then be that the simultaneous presence of (A) and (B) and the demonstration of a causal connection between these two for the Heisenberg case strongly suggests that this causal connection persists to this day as well. I will draw the conclusions from this assertion at the very end. But first, let us discuss the four central points:

A. S. Blum, *Heisenberg's 1958 Weltformel and the Roots of Post-Empirical Physics*, SpringerBriefs in History of Science and Technology,
https://doi.org/10.1007/978-3-030-20645-1_5

- Let us begin with the theory's genesis. We have seen how from a simple philosophical principle (reductive monism) emerged, almost algorithmically, the basic structure of Heisenberg's Hamiltonian. This imparted Heisenberg's theory (to him, at least) with a sense of inevitability that partially explains his persistence in pursuing it. The quasi-algorithmical derivation of Heisenberg's Hamiltonian was made possible by the stringent constraints imposed on theorizing by the demands of Lorentz covariance (i.e., special relativity) and of quantum theory. These are to a large extent the constraints that govern theorizing in particle physics to this day, and it is easy to see how, as in Heisenberg's case, the ease that they provide in model building can impart an unwarranted impression of necessity. And indeed, we find arguments for the inevitability of string theory, e.g., in the no-alternatives argument of Dawid (2013).[1]
  It should be noted, however, that these restrictions ultimately allow QFT to reach rather strong verdicts on philosophical principles that might at first glance appear far to vague to pin down. While Heisenberg himself never gave up on his nonlinear spinor theory, I think it is fair to say that its inability to ever be empirically adequate was after all generally and legitimately established. The case can then be made that the failure of Heisenberg's theory implies a refutation of monism, as long as one wants to hold on to basic tenets of relativistic QFT. Contemporary grand unified theories (the closest analogue to Heisenberg's theory, which for the most part had nothing to say on gravity) always make a distinction at least between fermions and gauge bosons, a distinction that Heisenberg unsuccessfully tried to overcome, and even supersymmetric theories need to distinguish between two kinds of super-multiplets, one type containing the gauge bosons, the other the matter fermions.
- Another feature of the genesis of Heisenberg's theory deserves mentioning and that is the element of frustration. As we have seen, Heisenberg turned to his foundational pursuits in high theory also because he was unable to pursue experimental nuclear and cosmic ray physics at his institute after World War II. The large scale necessary for experimentation in modern fundamental physics can lead to an inaccessibility of empirical data that in turn motivates non-empirical theory construction. We saw this on a local level in Heisenberg's case and see it repeated on a global level, as cutting-edge particle accelerator experiments now take decades to plan and construct, always in danger of losing funding. It would be a worthwhile endeavor to historically study the connection between the cancellation of the superconducting supercollider in 1993 and the rise of string theory, which is usually attributed to the second string theory revolution, a set of theoretical breakthroughs in the years 1994/95 (Rickles 2014, Chap. 10, in particular Fig. 10.1).
- Turning to the assessment of Heisenberg's theory, we have seen how hard it was to establish its consistency or inconsistency, a difficulty that can be immediately traced to the fact that this question was not settled even for the likes of QED. It is

[1] See also the reformulation of the no-alternatives argument as applied to loop quantum gravity by Smolin (2014).

only in this context that one can understand why Heisenberg (and for some time also Pauli) treated Heisenberg's victory in the battle of Ascona as a major breakthrough. All that Heisenberg had provided was a sketch of a proof that a theory with an indefinite metric could still in principle generate a unitary S-Matrix, i.e., he had refuted a possible a priori argument that might be wielded against his theory in lieu of an actual inconsistency proof. While this appears a rather feeble success, it must be realized that given the practical impossibility of a more detailed proof of inconsistency (and the difficulty of empirical refutation discussed in the next item), eliminating such simple a priori counterarguments (i.e., indefinite metrics generically violate unitarity) was really the only major hurdle a research program in (non-perturbative) QFT had to take, before it could be considered legitimate at least in principle.

QFT has certainly made great progress in the last 60 years, with regards to its mathematical and formal coherence. However, there are still no rigorous proofs for the "existence" (i.e., the mathematically consistent constructibility) of even the simplest interacting relativistic QFTs in four dimensions, and the debate on what QFT actually is, the unsuccessful attempts of the axiomatists or the hand-waving Lagrangians and path integrals of the phenomenologists, continued into the 21st Century (Wallace 2006; Fraser 2009). This lack of any sort of rigorous existence or consistency proof is true a fortiori for string theory. And here we observe, just as in Heisenberg's case, how the 1984 refutation of a possible a priori argument against string theory, the existence of anomalies, is hailed as a major breakthrough, going by the name of "first string revolution" (Rickles 2014, Chap. 8).

- The final feature we need to address is the great difficulty in extracting reliable empirical pre- and postdictions from a fundamentally non-perturbative quantum field theory such as Heisenberg's. We need not dwell here on Heisenberg's attempts to circumvent this difficulty with the now largely forgotten Tamm-Dancoff approximations. Instead we focus on the other method Heisenberg used, the short-circuiting of detailed dynamical calculations through the study of the symmetry properties of his Hamiltonian. We have seen how Heisenberg used these symmetry properties to argue that his monistic theory would be in principle able to reproduce the rich subatomic particle spectrum, even if there were no methods available (Tamm-Dancoff aside) to show how that spectrum actually arose from his theory dynamically. Heisenberg and Pauli were hardly alone in the late 1950s in placing an emphasis on symmetry, as witnessed, e.g., by the global symmetry models of Schwinger (1957) or Gell-Mann (1957), and in all cases we observe symmetries playing the dual role of providing heuristic input for theory construction, while also allowing for concrete empirical predictions to be made in absence of a method to solve the theory's equations.

  As with its mathematical coherence, QFT certainly has made major advances in the last 60 years as regards the pre- and postdictivity of non-perturbative quantum field theories, especially through the methods of lattice gauge theory (and thus through the increased power of machine computation). But while these methods fare well with the non-abelian gauge theory of quantum chromodynamics, they would essentially have to be reinvented for a theory such as Heisenberg's, which claimed to have solutions with a negative norm in Hilbert space and which had no safe perturbative regime from which to extrapolate (as does quantum chromodynamics, due to asymptotic freedom). The same holds true for any theory that breaks out of the standard model norm, such as, again, string theory. So again we can see that the issues that led Heisenberg to adapt symmetry as a substitute for dynamics are still with us.

As for the complementary statement, that symmetry still is an integral part of research in contemporary particle physics and quantum field theory, that is so obviously true that it hardly warrants mentioning. However, symmetry arguments are here hardly viewed as any kind of substitute; to the contrary, they are celebrated for their heuristic power in theory construction and their sheer mathematical beauty. Both of these factors certainly played a role for Heisenberg as well, but, as we have seen, his predilection for symmetry also stemmed from the fact that it allowed him to make statements about the empirical content of his theory, even as detailed dynamical calculations were out of the question. Given the persistence of this difficulty in contemporary QFT, it seems likely that this fact continues to motivate the prevalence of symmetry arguments, even though it is, as far as I know, never mentioned, not even in severe criticisms of the complementary use of symmetry arguments for theory construction (Hossenfelder 2018). This is all the more surprising as, if my own experience is any guide, the first experience that physics undergraduates make with symmetry arguments is when a tutor solves a homework problem in two lines instead of the three pages of detailed dynamical calculations that it took the student.

While no hard-and-fast list of the features of "post-empirical physics" exists, the features (A) identified in the above list appear to form a rather good starting point: A predilection for high theory; the construction of theoretical models with little new empirical input (what Galison 1995 in this context called "Theory Unbound"); a strong focus on questions internal to a given research program; and an emphasis on arguments of symmetry and elegance. The analysis of the Heisenberg story complements this list with a list of underlying causes, which should be studied in any further historico-philosophical analysis of the origins of the post-empirical turn of current fundamental physics, some obvious, some less intuitive. These are: the increasing difficulty of obtaining empirical input; the ease of highly constrained theory construction in modern physics; the unresolved foundational status of QFT; and the difficulty of extracting empirical predictions from non-perturbative QFT. I therefore think the relevance of the story of Heisenberg's *Weltformel*, beyond the immediate interest in the story itself, lies in the pointers it can give us for the later history of fundamental physics and the origins of its current crisis.

Are there also normative conclusions? Pauli himself drew a very strict normative conclusion from the Heisenberg episode. As he wrote to Walter Thirring on 20 May 1958: "In any case, one should, for a long time, give projects of theoreticians *no more advance credit*." It is left to the reader to assess whether this is too harsh.

References

Ascoli, R., & Heisenberg, W. (1957). Zur Quantentheorie nichtlinearer Wellengleichungen IV. Elektrodynamik. *Zeitschrift für Naturforschung, 12a*, 177–187.

Ballam, J., Fitch, V., Fulton, T., Huang, K., Rau, R. R., & Treiman, S. B. (Eds.). (1956). *High Energy Nuclear Physics: Proceedings of the Sixth Annual Rochester Conference, April 3–7, 1956*. New York: Interscience.

Bernstein, J. (1996). *Hitler's Uranium Club: The secret recordings at Farm Hall*. Woodbury, NY: American Institute of Physics.

Bernstein, J. (2018). *A bouquet of Dyson and other reflections on science and scientists*. World Scientific.

Blum, A. (2017). The state is not abolished, it withers away: How quantum field theory became a theory of scattering. *Studies in History and Philosophy of Modern Physics, 60*, 46–80.

Blum, W., Dürr, H.-P., Rechenberg, H., & (Eds.)., (1993). *Werner Heisenberg-Gesammelte Werke (Vol.* A III): Springer.

Borowitz, S., & Kohn, W. (1949). On the electromagnetic properties of nucleons. *Physical Review, 76*(6), 818–827.

Borrelli, A. (2015a). The making of an intrinsic property: "Symmetry heuristics" in early particle physics. *Studies in History and Philosophy of Science, 50*, 59–70.

Borrelli, A. (2015b). The story of the Higgs boson: The origin of mass in early particle physics. *European Physical Journal H, 40*, 1–52.

Carson, C. (2010). *Heisenberg in the atomic age*. Cambridge University Press.

Carson, C. L. (1995). *Particle physics and cultural politics: Werner Heisenberg and the shaping of a role for the physicist in postwar West Germany*. Ph.D. thesis, Harvard University.

Cassidy, D. C. (1981). Cosmic ray showers, high energy physics, and quantum field theories: Programmatic interactions in the 1930s. *Historical Studies in the Physical Sciences, 12*(1), 1–39.

Cassidy, D. C. (1992). *Uncertainty: The life and science of Werner Heisenberg*. New York: W. H. Freeman.

Chew, G. (1954). Renormalization of meson theory with a fixed extended source. *Physical Review, 94*(6), 1748–1754.

Cushing, J. T. (1990). *Theory construction and selection in modern physics: The S Matrix*. Cambridge: Cambridge University Press.

Dawid, R. (2013). *String theory and the scientific method*. Cambridge University Press.

Deltete, R. J. (1983). *The energetics controversy in late nineteenth-century Germany: Helm, Ostwald and their critics*. Ph.D. thesis, Yale University.

D'Espagnat, B., & Prentki, J. (1955). Possible mathematical formulation of the Gell-Mann model for new particles. *Physical Review, 99*(1), 328–329.

A. S. Blum, *Heisenberg's 1958 Weltformel and the Roots of Post-Empirical Physics*, SpringerBriefs in History of Science and Technology,
https://doi.org/10.1007/978-3-030-20645-1

Dürr, H. P. (1982). Heisenbergs einheitliche Feldtheorie der Elementarteilchen. *Nova Acta Leopoldina*, *55*(248), 93–136.
Dürr, H.-P., Heisenberg, W., Mitter, H., Schlieder, S., & Yamazaki, K. (1959). Zur Theorie der Elementarteilchen. *Zeitschrift für Naturforschung*, *14a*, 441–485.
Dyson, F. J. (1953). The wave function of a relativistic system. *Physical Review*, *91*(6), 1543–1550.
Eckert, M. (2000). Theorien, die die Welt bewegen: Wie die Physik zur "Jahrhundertwissenschaft" wurde. *Kultur & Technik*, *4*, 18–23.
Fässler, A., & Schmid, E. (2006). Nachruf Karl Wildermuth. *Physik Journal*, *5*(1), 53.
Fierz, M. (1950). Über die Bedeutung der Funktion D_c in der Quantentheorie der Wellenfelder. *Helvetica Physica Acta*, *23*, 731–739.
Forman, P. (1971). Weimar culture, causality, and quantum theory, 1918–1927: Adaptation by german physicists and mathematicians to a hostile intellectual environment. *Historical Studies in the Physical Sciences*, *3*, 1–115.
Fraser, D. (2009). Quantum field theory: Underdetermination, inconsistency, and idealization. *Philosophy of Science*, *76*(4), 536–567.
Galison, P. (1995). Theory bound and unbound: Superstrings and experiments. In F. Weinert (Ed.), *Laws of nature: Essays on the philosophical, scientific and historical dimensions* (pp. 369–408). (de Gruyter).
Gell-Mann, M. (1957). Model of the strong couplings. *Physical Review*, *106*(6), 1296–1300.
Grundgleichung der Materie. (1958). *Physikalische Blätter*, *14*(5), 238–239.
Gürsey, F. (1958). Relation of charge independence and baryon conservation to Pauli's transformation. *Il Nuovo Cimento*, *7*(3), 411–415.
Haag, R. (2010). Some people and some problems met in half a century of commitment to mathematical physics. *European Physical Journal H*, *35*(3), 263–307.
Heisenberg, W. (1931). Die Rolle der Unbestimmtheitsrelationen in der modernen Physik. *Monatshefte für Mathematik und Physik*, *38*, 365–372.
Heisenberg, W. (1933). Zur Geschichte der physikalischen Naturerklärung. *Bericht über die Verhandlungen der Sächsische Akademie der Wissenschaften zu Leipzig, mathematisch-physikalische Klasse*, *85*, 29–40.
Heisenberg, W. (1934). *Atomtheorie und Naturerkenntnis. Universitätsbund Göttingen. Mitteilungen*, *16*(1), 9–20.
Heisenberg, W. (1941). Die Goethe'sche und die Newton'sche Farbenlehre im Lichte der modernen Physik. *Geist der Zeit. Wesen und Gestalt der Völker (Hochschule im Ausland), Neue Folge*, *19*(5), 261–275.
Heisenberg, W. (1943a). Besprechung von "Gregor Wentzel-Einführung in die Quantentheorie der Wellenfelder". *Die Naturwissenschaften*, *31*(21/22), 251–252.
Heisenberg, W. (Ed.). (1943b). *Vorträge über kosmische Strahlung*. Springer.
Heisenberg, W. (1949). Die gegenwärtigen Grundprobleme der Atomphysik. In *Wandlungen in den Grundlagen der Naturwissenschaft* (8th ed., pp. 89–101). S. Hirzel, Zurich.
Heisenberg, W. (1950). Zur Quantentheorie der Elementarteilchen. *Zeitschrift für Naturforschung*, *5a*, 251–259.
Heisenberg, W. (1952). *Mesonerzeugung als Stoßwellenproblem. Zeitschrift für Physik*, *133*, 65–79.
Heisenberg, W. (Ed.). (1953a). *Kosmische Strahlung: Vorträge gehalten im Max-Planck-Institut für Physik*. Göttingen: Springer.
Heisenberg, W. (1953b). Zur Quantisierung nichtlinearer Gleichungen. In *Akademie der Wissenschaften in Göttingen, math.-phys- Klasse. Nachrichten* (pp. 111–127).
Heisenberg, W. (1954). Zur Quantentheorie nichtrenormierbarer Wellengleichungen. *Zeitschrift für Naturforschung*, *9a*, 292–303.
Heisenberg, W. (1956). Erweiterungen des Hilbert-Raums in der Quantentheorie der Wellenfelder. *Zeitschrift für Physik*, *144*, 1–8.
Heisenberg, W. (1957a). Lee model and quantisation of non linear field equations. *Nuclear Physics*, *4*, 532–563.

Heisenberg, W. (1957b). Quantum theory of fields and elementary particles. *Reviews of Modern Physics*, *29*(3), 269–278.

Heisenberg, W. (1958). Die Plancksche Entdeckung und die philosophischen Grundfragen der Atomlehre. *Naturwissenschaften*, *45*(10), 227–234.

Heisenberg, W. (1969). *Der Teil und das Ganze*. München: Piper.

Heisenberg, W. (1984). Erkenntnistheoretische Probleme in der modernen Physik. In W. Blum, H.-P. Dürr, & H. Rechenberg (Eds.), *Gesammelte Werke Abteilung C: Allgemeinverständliche Schriften* (Vol. 1, pp. 22–28). Piper: München and Zürich. (Physik und Erkenntnis (1927–1955)).

Heisenberg, W., Kortel, F., & Mitter, H. (1955). Zur Quantentheorie nichtlinearer Wellengleichungen III. *Zeitschrift für Naturforschung*, *10a*, 425–446.

Heisenberg, W., & Pomerans (translator), A. J. (1971). *Physics and beyond: Encounters and conversations*. World Perspectives, New York, Evanston, and London.

Heisenberg, W., & von Weizsäcker, C. (1948). Die Gestalt der Spiralnebel. *Zeitschrift für Physik*, *125*, 290–292.

Henning, E., & Kazemi, M. (2016). *Handbuch zur Institutsgeschichte der Kaiser-Wilhelm/Max-Planck-Gesellschaft zur Förderung der Wissenschaften (1911–2011)* (Vol. 2, pp. 1177–1216). Max-Planck-Gesellschaft.

Hoffmann, D. (1996). Wider die geistige Trennung: Die Max-Planck-Feier(n) in Berlin 1958. *Deutschland Archiv*, *29*, 525–534.

Hossenfelder, S. (2018). *Lost in math: How beauty leads physics astray*. Basic Books.

Huggett, N. (2014). Review of "String theory and the scientific method" by Richard Dawid. *Notre Dame Philosophical Reviews*.

Jordan, P. (1952). Über die Erhaltungsssätze der Physik. *Zeitschrift für Naturforschung*, *7*(1), 78–81.

Källén, G., & Pauli, W. (1955). On the mathematical structure of T. D. Lee's model of a renormalizable field theory. *Det Kgl. Danske Videnskabernes Selskab Mathematisk-fysiske Meddelelser*, *30*, 3–23.

Kamefuchi, S. (1951). Note on the direct interaction between spinor fields. *Progress of Theoretical Physics*, *6*(2), 175–181.

Kant, H. (1996). Albert Einstein, Max von Laue, Peter Debye und das Kaiser-Wilhelm-Institut für Physik in Berlin (1917–1939). In B. vom Brocke & H. Laitko (Eds.), *Die Kaiser-Wilhelm-/Max-Planck-Gesellschaft und ihre Institute* (pp. 227–243). Berlin: de Gruyter.

Karpman, V. (1957). On the theory of strange particles. *Journal of Experimental and Theoretical Physics (USSR)*, *32*, 939–940.

Lee, T. D. (1954). Some special examples in renormalizable field theory. *Physical Review*, *95*(5), 1329–1334.

Lehmann, H. (1954). Über Eigenschaften von Ausbreitungsfunktionen und Renormierungskonstanten quantisierter Felder. *Nuovo Cimento*, *11*, 342–357.

Low, F. (1956). Pi mesodynamics. *Reviews of Modern Physics*, *29*(2), 216–222.

Meier, C. (Ed.). (1992). *Wolfgang Pauli und C. G. Jung: Ein Briefwechsel (1932–1958)*. Springer.

Pauli, W. (1941). Relativistic field theories of elementary particles. *Reviews of Modern Physics*, *13*, 203–232.

Pauli, W. (1956). Die Wissenschaft und das äbendländische Denken. *Europa-Erbe und Aufgabe, Internationaler Gelehrtenkongress, Mainz, 1955* (pp. 71–79). Wiesbaden: F. Steiner Verlag.

Pauli, W. (1957). On the conservation of the lepton charge. *Nuovo Cimento*, *6*, 204–215.

Pauli, W., & Villars, F. (1949). On the invariant regularization in relativistic quantum theory. *Reviews of Modern Physics*, *21*, 434–444.

Rechenberg, H. (1989). The early S-matrix theory and its propagation (1942–1952). In L. M. Brown, M. Dresden, & L. Hoddeson (Eds.), *Pions to quarks*. Cambridge: Cambridge University Press.

Rechenberg, H. (1996). Werner Heisenberg und das Forschungsprogramm des Kaiser-Wilhelm-Instituts für Physik. In B. vom Brocke & H. Laitko (Eds.), *Die Kaiser-Wilhelm-/Max-Planck-Gesellschaft und ihre Institute* (pp. 245–262). Berlin: de Gruyter.

Rettig, A. (2014). *Halbgenie und Viertelfaust: Heisenbergs und Paulis Quantenfeldtheorie von 1958*. Master's thesis, Universität Regensburg.

Rickles, D. (2014). *A brief history of string theory*. Springer.

Schwinger, J. (1948). Quantum electrodynamics. *I. A covariant formulation. Physical Review, 74*(10), 1439–1461.

Schwinger, J. (1949). Quantum electrodynamics. II. *Vacuum polarization and self-energy. Physical Review, 75*, 651–679.

Schwinger, J. (1957). A theory of the fundamental interactions. *Annals of Physics, 2*, 407–434.

Smolin, L. (2006). *The trouble with physics: The rise of string theory, the fall of a science, and what comes next*. Houghton Mifflin Harcourt.

Smolin, L. (2014). Review of "String theory and the scientific method" by Richard Dawid. *American Journal of Physics, 82*, 1105–1107.

Stoelzner, M. (2001). Opportunistic axiomatics-Von Neumann on the methodology of mathematical physics. In M. Rédei & M. Stoelzner (Eds.), *John von Neumann and the Foundations of Quantum Physics* (pp. 35–62). Dordrecht: Springer Science+Business Media.

Thirring, W. E. (1958). A soluble relativistic field theory. *Annals of Physics, 3*(1), 91–112.

von Meyenn, K. (Ed.). (1993). *Wolfgang Pauli: Wissenschaftlicher Briefwechsel mit Bohr, Einstein, Heisenberg u.a.* (Vol. III). Springer, Berlin. (1940–1949).

von Meyenn, K. (ed.). (1996). *Wolfgang Pauli: Wissenschaftlicher Briefwechsel mit Bohr, Einstein, Heisenberg u.a.* (Vol. IV). Springer, Berlin. (Part I: 1950–1952)

von Meyenn, K. (Ed.). (1999). *Wolfgang Pauli: Wissenschaftlicher Briefwechsel mit Bohr, Einstein, Heisenberg u.a.* (Vol. IV). Springer, Berlin. (Part II: 1953–1954).

von Meyenn, K. (Ed.). (2001). *Wolfgang Pauli: Wissenschaftlicher Briefwechsel mit Bohr, Einstein, Heisenberg u.a.* (Vol. IV). Springer, Berlin. (Part III: 1955–1956).

von Meyenn, K. (Ed.). (2005). *Wolfgang Pauli: Wissenschaftlicher Briefwechsel mit Bohr, Einstein, Heisenberg u.a.* (Vol. IV). Springer, Berlin. (Part IV: 1957–1958).

Walker, M. (1992). Physics and propaganda: Werner Heisenberg's foreign lectures under national socialism. *Historical Studies in the Physical and Biological Sciences, 22*(2), 339–389.

Wallace, D. (2006). In defence of Naiveté: The conceptual status of Lagrangian quantum field theory. *Synthese, 151*(1), 33–80.

Wigner, E. P. (1949). Invariance in physical theory. *Proceedings of the American Philosophical Society, 93*(7), 521–526.

Woit, P. (2006). *Not even wrong: The failure of string theory and the search for unity in physical law*. Basic Books.

Wüthrich, A. (2013). Feynman's struggle and Dyson's surprise: The development and early application of a new means of representation. In S. Katzir, C. Lehner, & J. Renn (Eds.), *Traditions and transformations in the history of quantum physics* (Vol. 5, pp. 271–289). Max Planck Research Library for the History and Development of Knowledge Proceedings Edition Open Access: Berlin.

Zimmermann, W. (2001). Harry Lehmann, der Feldverein und die Anfänge der axiomatischen Quantenfeldtheorie. In K. von Meyenn (Ed.), *Wolfgang Pauli: Wissenschaftlicher Briefwechsel mit Bohr, Einstein, Heisenberg u.a. Band IV, Teil III: 1955–1956* (pp. 68–74). Springer.

Printed in the United States
By Bookmasters